FATAL
POLITICS

KEN HUGHES

FATAL POLITICS

THE NIXON TAPES, THE VIETNAM WAR,
AND THE CASUALTIES OF REELECTION

University of Virginia Press
CHARLOTTESVILLE AND LONDON

To Kenneth J. Hughes Sr., Private, US Army
and Maryann Hughes

University of Virginia Press
© 2015 by the Rector and Visitors of the University of Virginia, Miller Center
All rights reserved
Printed in the United States of America on acid-free paper

First published 2015

ISBN 978-0-8139-3802-8 (cloth)
ISBN 978-0-8139-3803-5 (e-book)

9 8 7 6 5 4 3 2 1

Library of Congress Cataloging-in-Publication Data is available from the Library of Congress.

Read and listen to the presidential conversations online at www.fatal-politics.org

And let all sleep; while to my shame I see
The imminent death of twenty thousand men
That for a fantasy and trick of fame
Go to their graves like beds . . .
        —WILLIAM SHAKESPEARE, *HAMLET*

Will it make it easier on you
Now you got someone to blame?
        —U2, "ONE"

CONTENTS

Introduction ix

Fatal Politics 1

Vietnamization 11

"A Nightmare of Recrimination" 16

"A Hell of a Shift" 20

How to Kill a Withdrawal Deadline 23

"I'm Being Perfectly Cynical" 26

"We Want a Decent Interval" 30

Meeting Zhou 34

"Old Friends" 36

"He Deserves Our Confidence" 37

JFK v. Nixon 39

The Kennedy Critique 45

The Liberal Mistake 49

"Super Secret Agent" 54

Sixty-Six Percent for Six Months 55

"One Arm Tied Behind" 64

"Why Does the Air Force Constantly Undercut Us?" 67

The Appearance of Success 69

"Any Means Necessary" 70

"A Russian Game, a Chinese Game and an Election Game" 75

"It Could Be a Bit Longer" 78

The Democrats 81

"No One Will Give a Damn" 83

"Idealism with Integrity" 86

"Our Terms Will Eventually Destroy Him" 89

Blowup 1968 97

"We're behind the Trees!" 98

"Saving Face or Saving Lives" 101

"Brutalize Him" 103

Kissinger v. Thieu 111

"No Possibility Whatever" 116

"The Man Who Should Cry Is I" 117

"The Fellow Is Off His Head" 119

No Coalition Government 126

"Peace Is at Hand" 128

"A Little Bit Diabolically" 135

The Chennault Affair 140

"The Clearest Choice" 141

Election Day 1972 143

Promises and Threats 147

Christmas Bombing 150

"Let Us Be Proud" 159

The Prisoners Dilemma 161

The Final Cutoff 163

Stabbed in the Back 165

"We Can Blame Them for the Whole Thing" 167

Nixon's *Dolchstoßlegende* 170

Unearthing Nixon's Strategy 173

The University of Virginia's Miller Center 179

Decision Points 181

The Nixon Tapes 185

Interpretive Inertia 187

Last Days in Vietnam 192
A Better War 197
The Aid-Cutoff Myth 200
How Wars Don't End 202
Questions Unasked 206

Acknowledgments 211
Notes 215
Index 255

INTRODUCTION

"No event in American history is more misunderstood than the Vietnam War." No one did more to keep it that way than the author of the preceding sentence, Richard Milhous Nixon. It's the opening line of his 1985 best seller, *No More Vietnams*.[1] In the former president's version of events, he won the war, only to watch helplessly as Congress "proceeded to snatch defeat from the jaws of victory."[2] One reviewer wrote that Nixon had fabricated a "stabbed in the back" myth, one all too reminiscent of the *Dolchstoßlegende* that German militarists created to blame their defeat in World War I on their country's civilians (rather than on the predictable shift in the balance of battlefield power caused by America's entrance into the war on the Allied side).[3] In the battle over history, however, Nixon had a crucial advantage over his critics. They didn't have access to the best evidence: the classified record of Nixon's foreign policy making, especially his secretly recorded White House tapes. The tapes covered the critical period—February 16, 1971, to July 12, 1973—when Nixon withdrew the last American troops from Vietnam, negotiated a settlement of the war with North Vietnam's Communist government, engineered the diplomatic opening to China, established a détente with Russia, and won a landslide reelection. The tapes reveal the complex and subtle interplay between all these actions—including the ways Nixon manipulated geopolitical events for domestic political gain. Unsurprisingly, he fought until his death in 1994 to keep the American people from hearing the tapes; tragically, it took the federal government nearly two decades, until 2013, to finish declassifying these invaluable and (thanks to Nixon's sound-activated recording system) comprehensive historical records. By then it was almost too late. Politicians, policy makers, and pundits now routinely invoke Nixon's backstabbing myth as reason to block attempts to end America's twenty-first-century wars in Iraq and Afghanistan. They even promote Nixon-era strategy as a path to victory in both these countries.

Nixon's tapes reveal, however, that he merely came up with a politically acceptable substitute for victory. Neither Nixon nor any of his military or civilian advisers ever devised a workable strategy to win the war, but Nixon and National Security Adviser Henry Kissinger devised a brilliant, if ruthless, strategy to win the election. As Nixon and Kissinger saw it, victory in the presidential election did not depend on victory in Vietnam; they merely had to postpone—not prevent—the Communist takeover of South Vietnam until sometime after November 1972. To accomplish this end, Nixon kept American soldiers in Vietnam into the fourth year of his presidency, at the cost of thousands of American lives.

That was the military side of his secret strategy. Like his official, publicly announced strategy, Nixon's secret strategy had both a military and a diplomatic side. Officially, Nixon's strategy was "Vietnamization and negotiation." Publicly, Nixon said Vietnamization would train and equip the South Vietnamese to defend themselves without the need for American troops. Secretly, Nixon used Vietnamization as an excuse to prolong the war long enough to delay South Vietnam's fall past Election Day 1972.

As for the diplomatic side, Nixon said publicly that the aim of negotiations was to reach an agreement with the North guaranteeing the South's right to choose its government by free elections. Secretly, however, he did not require the North to abandon its goal of military conquest of the South. Instead, he settled for a "decent interval"—a period of a year or two—between his final withdrawal of American troops and the Communists' final takeover of South Vietnam. For Nixon to completely evade the blame for defeat, he had to do more than prop up the Saigon government through 1972. If it fell shortly after he brought the troops home, Americans would see that their soldiers had died in vain, and Nixon would go down in history as the first president to lose a war. Nixon could avoid this fate, however, if the Communists gave him a "decent interval."

Nixon and Kissinger's secret strategy, though clearly immoral, did not spring simply from the character flaws of two men.[4] It was a logical, if extreme, outgrowth of Cold War politics. Successful Cold War politicians blamed their opponents for losing countries to the Communists—even countries where Americans and Communists had not been fighting. Republicans gained control of both houses of Congress in 1946 in part by blaming Democrats for "losing" Eastern Europe to Communism; the GOP also picked up House and Senate seats in 1950 by blaming Democrats for "losing" China to Mao Zedong's Communist revolutionaries. JFK won the presidency in 1960 in part by blaming Republicans for "losing" Cuba to Communist Fidel Castro. Given this political tendency, Nixon had reason to fear that if he lost Vietnam in his first term, American voters would deny him a second one. That was political reality. It doesn't excuse what Nixon and Kissinger did; it merely shows that they acted on the basis of rational political calculation.

Just as important as how Nixon won the 1972 election is how his opponent lost. According to the Emory University professor Drew Westen, political campaigns are built, in part, on "the story your opponent is telling about himself" and "the story you are telling about your opponent."[5] The story Nixon told about himself—that he would keep the war going only until South Vietnam could defend itself or the North agreed to let it choose its government by free elections—was not true. Unfortunately, the story his opponent told about him wasn't true, either.

The Democratic presidential nominee, Sen. George McGovern of South Dakota, did not accuse Nixon of prolonging the war and faking peace for political gain, of putting his reelection campaign above the lives of American soldiers, of sacrificing them for a fig leaf behind which he would secretly surrender the South to the Communists. Instead, McGovern and other liberals claimed that Nixon would never allow Saigon to fall, that a vote to reelect the president was a vote for four more years of war. Although Nixon's tapes show that he was not the steadfast ally of Saigon that he pretended to be, McGovern's charges counterproductively reinforced the image that Nixon had carefully cultivated.

Worse, McGovern didn't know what to do in October 1972 when South Vietnamese president Nguyen Van Thieu publicly declared that the deal Nixon and Kissinger made with North Vietnam was a sellout and surrender to the Communists. The charge was both damning and true, but because it contradicted what McGovern had been saying, he failed to seize the opportunity it offered. Autopsies of McGovern's campaign usually detail his parade of political pratfalls through the summer and fall of 1972, but they neglect its central strategic flaw. Vietnam was the biggest issue of the general election campaign, and McGovern fumbled it.

It didn't have to be that way. Sen. Ted Kennedy, D-Massachusetts, provided an alternative strategy when he accurately accused Nixon of cynically using Vietnamization as a fraudulent cover for timing military withdrawal to his reelection campaign. Kennedy's brother-in-law Sargent Shriver—McGovern's running mate in the fall campaign—accurately accused Nixon of negotiating surrender. If McGovern had told *that* story throughout the campaign, not only would he have been right, but Nixon's troop withdrawals and Thieu's blowup would have provided the story with credible confirmation. Instead, McGovern and other liberals lost control of the foreign policy narrative, telling a story about Nixon that was both false and flattering to the man it was designed to defeat.

Though the 1972 campaign is decades past, the myth Nixon made (and McGovern unwittingly reinforced) persists, now hallowed as if it were settled history. Right and Left agree that Nixon was determined to use American military power to preserve South Vietnam until Congress tied his hands. The Right calls that losing Vietnam, the Left calls it ending the war, but both agree that's what happened. As we shall see, Nixon invited Congress to pass legislation denying him authority to militarily intervene in Vietnam, despite having the votes he needed to sustain a veto. Legislation barring military intervention throughout Indochina (Vietnam, Laos, and Cambodia) passed with a veto-proof majority only after Nixon's conservative supporters joined his liberal opponents and accepted his invitation. Like Nixon's secret military and diplomatic strategy, this legislative maneuver enabled him to deny responsibility for losing Vietnam.

The danger of the backstabbing myth that Nixon spun in *No More Vietnams* is that it paves the way for more Vietnams. Today, no politician wants to be accused of losing Iraq or Afghanistan. That doesn't mean any of them (or their civilian and military advisers) have ever come up with a way to win either war. Nixonian myth, however, gives them a politically acceptable alternative to admitting failure. They can hold up the false hopes that training and equipping the local armies will enable them to replace American soldiers and that a political settlement will reconcile parties who have demonstrated their inclination to fight out their differences. The cost of false hope is measured in lost and shattered lives.

Fortunately, Nixon's myth is shattered by the evidence on Nixon's tapes and in his White House documents. I've been studying Nixon's tapes for decades, first as a journalist writing in the pages of the *New York Times Magazine, Washington Post, Boston Globe Magazine,* and other publications in the 1990s, and since 2000 as a researcher with the Presidential Recordings Program of the University of Virginia's Miller Center. The tapes tell a story that is both true and, potentially, lifesaving. Once we remember this hidden part of our past, we will no longer be condemned to repeat it.

FATAL
POLITICS

A PRIL 7, 1971, 8:58 p.m., the Oval Office. "Two minutes away from airtime. I'll cue you."

"Sure."

"I hope that you will favor this microphone." The floor director meant the one his crew had placed on the immaculate surface of the president's desk, not the five hidden beneath it. Only one man in the room knew about those.

President Richard M. Nixon said, "I'm not going to turn." Tonight would be the first television broadcast from the Oval Office since he began taping his conversations, secretly and automatically, seven weeks earlier. Secret Service technicians had proved to be very skilled at the kind of electronic surveillance they were trained to foil: the bugging of the president. They had done it reluctantly, on Nixon's order.[1] They placed the mics in the desk's open kneehole—not quite out of sight, but not where anyone would ordinarily look.

"Just be sure if you're going to talk, you address the microphone," said the director.

The president paused. *If* he was going to talk? He had requested airtime on all three television networks. Six months had passed since his last major address about Vietnam. In that time, he had made his biggest decision on the war. He had also made his biggest decision about his reelection campaign. The two were, in fact, one decision. Not that he was going to announce it. He couldn't risk the loss of everything he had worked to achieve. No, tonight he was going to announce an entirely different decision. It was designed to conceal—and buy time for him to implement—the more important, secret decision. Months of planning, weeks of writing, and hours of rehearsal had gone into this speech, including its seemingly ad lib conclusion. Nixon had even taken the trouble to write a fake ending for the speech that was included in the advance copies distributed to the press, just so the real one would come as a surprise.

The president exhaled and said, "That's pretty good advice."

Silence in the Oval Office.

"Oh, wait a minute," said Nixon. "Will it be on their screen if I leave this little handkerchief up here on this . . . ?" He was trying to hide the white cloth behind the squat, gray microphone stand on his desk. He gave no explanation; none was needed. Ever since his first presidential debate with John F. Kennedy, everyone in America knew about his problem. Now, before he spoke on TV, he had a makeup artist routinely apply antiperspirant above his mouth.[2]

"Let me check that," said the director.

"Oh, yeah, sure, but here's the thing—"

"Can you move a little to the left now?" asked some other TV guy.

The president tried. "Damn chair isn't . . . that all right?"

"Just a little, tiny bit more, sir," said the director.

"That better?" asked the president.

He got no answer to this or to the handkerchief question.

"He's on live in seven seconds."[3]

The Seal of the President faded in on CBS, NBC, and ABC, then faded out to reveal President Nixon seated between two flags. His desk was perfectly clean except for the typewritten speech, microphone stand, and something tiny and white sticking out behind the latter.

"Good evening, my fellow Americans. Over the past several weeks, you have heard a number of reports on TV, radio and in your newspapers on the situation in Southeast Asia. I think the time has come for me as president and commander in chief of our armed forces to put these reports in perspective, to lay all the pertinent facts before you and to let you judge for yourselves as to the success or failure of our policy. I am glad to be able to begin my report tonight by announcing that I have decided to increase the rate of American withdrawals for the period from May 1 to December 1[, 1971]."[4]

This wasn't the speech Nixon had originally intended to make. Four months earlier, in the weeks before Christmas, he had started making plans to announce the end of the war. All American troops home by December 31, 1971—that's what the people wanted (73 percent of them, according to Gallup).[5]

Henry Kissinger had talked him out of it. No one knew why Nixon had picked him to be national security adviser. It wasn't that Kissinger lacked intellectual stature (as author of the best-selling *Nuclear Weapons and Foreign Policy*) or academic credentials (as a professor of government at Harvard) or a compelling life story (as a Jewish refugee from the Nazis who, as a sergeant in the US Army, took part in overthrowing Hitler's regime). It was that Nixon hated Jews, intellectuals, and the Ivy League, especially Harvard. He made exceptions, however, for Jews, intellectuals, and Ivy Leaguers who labored loyally in his service. And

Kissinger did so with energy and skill. Through Kissinger, Nixon concentrated power over foreign policy in the White House, shrinking the roles of his secretaries of state and defense and their vast departmental bureaucracies, which were answerable to Congress and filled with employees who owed him no political loyalty. In no time at all, he made Kissinger the most powerful national security adviser in White House history.

Kissinger did not gain or wield that power by needlessly confronting the man who bestowed it, so at first he did not tell the president that he opposed a December 31, 1971, withdrawal deadline. Kissinger confided his reservations to the one Nixon aide arguably more powerful than himself: White House Chief of Staff H. R. Haldeman. Kissinger thought his name was Robert, since Nixon called him "Bob," and the never-ending train of memoranda to and from his office bore only his initials (which stood for Harry Robbins).[6] Haldeman was a meticulous, some might say compulsive, record keeper—taking down presidential orders on a succession of yellow pads, converting them into cartons of memos, even capturing presidential activities on Super 8 film (entire speeches without any sound).[7] It was his idea to make Nixon's secret taping system sound-activated. (He was the only member of the president's inner circle who knew that a reel-to-reel recorder started spinning in the White House basement whenever any of them entered the Oval Office.)[8] His self-discipline was legendary. At night, unknown to anyone in the White House, when Haldeman went home, he switched on a portable tape machine and recorded an entry in his diary.[9] He summarized Kissinger's objections to bringing the troops home by the end of 1971:

> H. R. "Bob" Haldeman: He thinks that any pull-out next year would be a serious mistake, because the adverse reaction to it could set in well before the '72 elections. He favors, instead, a continued winding-down and then a pull-out right at the fall of '72, so that if any bad results follow, they'll be too late to affect the election.[10]

The problem was simple: If Nixon withdrew the last American troops by December 31, 1971, South Vietnam might fall to the Communists in 1972. Then he would have to run for reelection as the first American president to have lost a war.

The national security adviser proposed a solution that was brilliant in its simplicity: delay the departure of the last American troops until after the election. If Saigon fell then, it would be too late for voters to hold the president (and his men) accountable.

Kissinger's logic was airtight. If Nixon lost the war, he'd lose the election—that is, if he lost the war *before* the election. It was a political paradox: the president had more to lose by doing the popular thing (bringing American soldiers home by the end of 1971) than by doing the unpopular thing (keeping them fighting and dying in Vietnam into 1972). "Seems to make sense," Haldeman told his diary.[11]

On December 21, 1970, the Monday before Christmas, Nixon had two historic meetings in the Oval Office. One became famous. Someone dropped off a handwritten letter to the president that morning at the Northwest Gate of the White House. On the stationery of American Airlines, Elvis Presley declared himself the president's admirer and volunteered "to help the country out" regarding the "drug culture." All he asked in return was that Nixon make him "a Federal Agent at Large." A White House aide quickly typed up a meeting request memo for Haldeman, who objected to one line: "If the president wants to meet with some bright young people outside of the government, Presley might be a perfect one to start with." This suggestion, unlike Kissinger's, was too much for Haldeman, who scribbled in the margin: "You must be kidding." He okayed the meeting anyway. You didn't have to be a former executive with the J. Walter Thompson ad agency (as Haldeman was) to see that a photo of Elvis and Nixon was manna from PR heaven.

At half past noon, an aide escorted Elvis into the Oval Office. As the president and "the King" shook hands, the White House photographer took more than the usual number of pictures. A prearranged discussion of Mr. Presley's ability to reach young people with an antidrug message ensued. "I can go right into a group of hippies and young people and be accepted," said Mr. Presley, according to the White House aide's careful notes.

"Well, that's fine," said the president. "But just be sure you don't lose your credibility." Elvis Presley's drug problems were not well known at the time; nor were they well hidden. Nixon knew how to read an audience.

So did Elvis, who fired a perfectly Nixonian blast at his only real rivals: "The Beatles, I think, are kind of anti-American." (They opposed the Vietnam War.) The president instructed an aide to fill Mr. Presley's request for a badge from the Bureau of Narcotics and Dangerous Drugs.[12] December 21, 1970, is still known as The Day Elvis Met Nixon; a picture commemorating the summit is the single most requested item in the National Archives.[13]

Nothing commemorates Nixon's other historic meeting of the day except an entry in Haldeman's tape-recorded diary. Alone with Haldeman and Kissinger, the president elaborated on his plans to bring the troops home from Vietnam by the end of 1971. He had decided to make the big announcement in April. Kissinger objected, this time directly to Nixon.

> H. R. "Bob" Haldeman: Henry argues against a commitment that early to withdraw all combat troops, because he feels that if we pull them out by the end of '71, trouble can start mounting in '72 that we won't be able to deal with, and which we'll have to answer for at the elections. He prefers instead a commitment to have them all out by the end of '72, so that we won't have to deliver finally until after the elections and therefore can keep our flanks protected. This would certainly seem to make more sense, and the president seemed to agree in general, but wants Henry to work up plans on it. He still feels he's got to make a major move in early '71.[14]

The analysis Kissinger gave the president was purely political. There is no indication in Haldeman's diary that other questions were answered or even raised: Would the additional year of war make South Vietnam capable of withstanding a Communist takeover after the final American withdrawal? How many more lives would be lost in that year? Was there anything wrong with timing withdrawal from a war for the benefit of an election campaign?

NBC's *Today Show* had aired a Barbara Walters interview with Kissinger that very morning. "You have said that the acid test of a policy is whether it is accepted by the public," said Walters.

"Yes."

"Does that mean that you try to find policies that the public will accept and reelect you for? Or do you try to tell the public what you think is right even though it may not be something too popular at the time?"

Seated before a bust of Abraham Lincoln, Kissinger scratched his nose with his thumb and replied with a self-deprecating quip: "I would think that any president who took my advice on public opinion would be in bad shape." He smiled.[15]

The president took his advice. He still made a big Vietnam speech in April. Instead of announcing the complete withdrawal of all American troops by the end of 1971, however, Nixon attacked the very idea. "Let me turn now to a proposal which at first glance has a great deal of popular appeal," the president said. The camera zoomed in. "If our goal is a total withdrawal of all our forces, why don't I announce a date now for ending our involvement? Well, the difficulty in making such an announcement to the American people is that I would also be making that announcement to the enemy. And it would serve the enemy's purpose and not our own. If the United States should announce that we will quit regardless of what the enemy does, we would have thrown away our principal bargaining counter to win the release of American prisoners of war."[16]

Nixon had elevated the plight of American POWs to national and international prominence. Right after his election as president in 1968, Sybil Stockdale, whose Navy pilot husband was shot out of the sky and captured by North Vietnam in 1965, enlisted the aid of other POW wives and a local newspaper to launch a telegram campaign aimed at bringing the POWs to the president-elect's attention. By the time Nixon arrived in the White House, two thousand telegrams greeted him.[17] They got action. The new secretary of defense, Melvin Laird, held a press conference calling on Hanoi to free the POWs.[18] The new secretary of state, William Rogers, condemned Hanoi's violations of the Geneva Conventions.[19] Defense and State dispatched public affairs teams to brief POW families across the country and get their stories into newspapers and on TV.[20] "The government has increased its public pressure to obtain the release of these prisoners," Richard Capen, assistant secretary of defense for public affairs, told one group.[21] The heartrending stories of the POW families were part of that public pressure. "Doubt Rules Lives of POW Families" was the *Los Angeles Times* headline over a picture of a young wife sitting alone in her backyard, gazing at a framed photo of her husband smiling in his Marine Corps uniform. The caption: "Living in Limbo—Wives and families of servicemen held by North Vietnam cannot be sure—are their men alive or dead?" Of her five sons, ranging in age from six to fourteen, Linda Morris said, "There's no doubt in their mind that their father will come home."[22] The courage, the absence of complaint, the simple faith, all would make a stone weep.

On the first Fourth of July of Nixon's presidency, Hanoi announced the release of three American POWs as a gesture.[23] It backfired when the former POWs revealed the torture prisoners endured. Navy Lieut. Robert Frishman told reporters that the North Vietnamese had left shrapnel in his right arm and allowed the wounds to fester. He denounced even more sharply the mistreatment of Lt. Comdr. John McCain III. After an anti-aircraft missile destroyed the wing of McCain's Skyhawk dive bomber, fracturing bones in his right leg and both of his arms, the North Vietnamese denied him medical treatment and locked him in solitary confinement.[24] President Nixon condemned the North for violating the Geneva Conventions on prisoners of war.[25]

The attention appeared to get some results. Having refused to give the names of the prisoners it held (another violation of the Geneva Conventions), Hanoi now began to release partial lists through American antiwar activists.[26] Hanoi Radio announced that families could send Christmas packages to the POWs in 1969. The Pentagon reminded North Vietnam and the world that the Geneva Conventions require regular exchange of mail and gifts.[27]

All this happened before the president secretly decided to time American military withdrawal from Vietnam to his 1972 reelection campaign. The decision to prolong the war was, in effect, a decision to prolong the POWs' captivity.

In public statements and private negotiations and even in secret talks, Hanoi made the price of the POWs' freedom plain: total American military withdrawal from Vietnam. The North Vietnamese thought they could overthrow the Saigon government once the Americans were out of the way. This made complete American withdrawal one of their basic war aims. While Nixon publicly claimed that keeping American soldiers in Vietnam gave Hanoi an incentive to free the prisoners, it actually gave Hanoi an incentive to hold them.

Nixon understood this. Weeks before his television speech, alone with Kissinger in the Oval Office on March 19, 1971, the president considered abandoning their election-oriented withdrawal timetable as a way to hasten the prisoners' release.

President Nixon: Henry, I've never been much for negotiation, but I think when we finally get down to the nut-cutting, it's very much to their advantage to have a negotiation to get us the hell out and give us those prisoners.
Kissinger: That's right. That's why—
President Nixon: And we've got to do it. And, you know, if they—if they'll make that kind of a deal, we'll make that any time they're ready.[28]

The national security adviser once more reminded the president of the political penalty for early withdrawal. Bringing the troops home too soon might cause Saigon to collapse at a time when voters could still hold him accountable.

Kissinger: Well, we've got to get enough time to get out. It's got to be because—
President Nixon: Oh, I understand.
Kissinger: —we have to make sure that they don't—
President Nixon: [*speaking over Kissinger*] I don't mean [*unclear*]
Kissinger: —knock the whole place over.

President Nixon: What?

Kissinger: Our problem is that if we get out after all the suffering we've gone through—

President Nixon: And then have it knocked over. Oh, I think [*unclear*]—

Kissinger: We can't have it knocked over—brutally—to put it brutally—before the election.

President Nixon: That's right.[29]

In this conversation, captured on tape (unbeknownst to Kissinger), neither man mentioned the additional losses that prolonging the war would cause—there is nothing about what it meant to American POWs in the North, American soldiers in the South, or the Vietnamese on either side. It was a subject the two men in the Oval Office habitually avoided.

On camera, the president shook the index finger of his right hand for emphasis as he said, "It is time for Hanoi to end the barbaric use of our prisoners as negotiating pawns and to join us in a humane act that will free their men as well as ours."[30] Both sides were using the prisoners as pawns—Hanoi to win the war, Nixon to win reelection. Sybil Stockdale and her fellow POW wives had turned to him for help; now they were helping him in ways they didn't know.

Casting the war as necessary to free the POWs was politically shrewd. Even failed attempts to free the prisoners were popular, as Nixon learned when he ordered an airborne Special Forces raid on Son Tay prison, just twenty miles from Hanoi. The "operation went beautifully," Kissinger reported to the president on November 21, 1970, "except no one was there." North Vietnam had moved all the POWs to a different prison camp three months earlier.[31] The commandos managed to get in and out without any American casualties. "They killed five guards," said Kissinger.[32]

"Even if I had known when the operation was being planned that the reports were out of date," Nixon later wrote, "I believe I would still have given my approval."[33] He had compelling reason to do so. Even after the public learned that the Special Forces had found no POWs at Son Tay, White House pollsters said support for future rescue attempts ran 73–8 percent in favor. The Son Tay raid itself, fruitless as it was, got 75–16 percent approval. When Nixon learned of those results, he responded with two words: "Jesus Christ."[34]

The president had the leader of the raid, Colonel "Bo" Gritz, over to the White House, and for an hour after their meeting he couldn't stop talking about how impressive Gritz was. Nixon was "as cranked up as I've ever seen him," Haldeman told his diary.[35] (Gritz later claimed, not implausibly, to be the inspiration for Rambo, the cinematic hero of modern rescue narratives.[36] *Rambo: First Blood Part II* echoed the popular plot of the Son Tay raid but fixed the ending.)

There was one conspicuous problem with Nixon's claim that he needed to keep troops in Vietnam as an incentive for the North to free the POWs: he was constantly shrinking the incentive. As he reminded television viewers, there were 540,000 American soldiers in Vietnam when he took office: "In June of 1969, I announced a withdrawal of 25,000 men; in September, 40,000; December, 50,000; April of 1970, 150,000. By the first of next month, May 1, we will have brought home more than 265,000 Americans—almost half of the troops in Vietnam when I took office."[37]

At a briefing the president gave senators at the White House before the speech, one of them had raised the obvious question. If the presence of more than a half million American soldiers in Vietnam had not given Hanoi a strong enough incentive to free the POWs, why would Hanoi free them when Nixon reduced the number to, say, 50,000?

"Of course," Nixon told Kissinger later, "I couldn't say to him, 'Look, when we get down to 50,000, then we'll make a straight-out trade—50,000 for the prisoner of wars—and they'll do it in a minute 'cause they want to get our ass out of there.'"

"That's right," said Kissinger.

The president laughed. "You know? Jesus!"[38]

Publicly, Nixon portrayed the release of the POWs as one of the aims of the war. Privately, he didn't see how he could possibly fail to achieve it. Hanoi would agree to release the POWs when he agreed to withdraw the troops. History was on his side. When Ho Chi Minh's forces overthrew the French colonial government of Vietnam in 1954, the Communists running the new government of North Vietnam had been delighted to send back home all the French POWs they had captured over the years.

The POW wives would get their husbands back eventually, but only when it was politically safe for Nixon to bring home the troops.

In the meantime, he took advantage of the manifest popular support for the POWs. Americans bought 50 million POW/MIA bumper stickers and 135 million POW/MIA postage stamps. The National League of Families of American Prisoners and Missing in Southeast Asia brought out an iconic flag, striking in stark black and white, that showed a prisoner's silhouette juxtaposed before a guard tower, with the words "You Are Not Forgotten." Soon it was flying all across America; in some places it still is. Copper bracelets bearing the names of imprisoned or missing Americans eventually sold sixty thousand per day. Ronald Reagan, the conservative Republican governor of California, wore one, as did the liberal Sen. George McGovern of South Dakota.[39]

McGovern bought the bracelet, but not Nixon's argument. As the first announced candidate for the Democratic presidential nomination—and the chief sponsor of legislation to set a December 31, 1971, withdrawal deadline—McGovern said the only way to get the POWs released was to set a date for every last American soldier to come home from Vietnam. A World War II bomber pilot and a professor of history, McGovern somehow managed to sound like a farmer, a liberal intellectual, and a biblical prophet all at once. A sharp nose and ascending forehead made him look like a hawk, but he was one of the Senate's first Vietnam doves. He had originally proposed that Congress force Nixon to bring the troops home by the end of 1970. On the day of the Senate vote, McGovern rose to the floor and said: "In one sense, this chamber reeks of blood. Every senator here is partly responsible for that human wreckage at Walter Reed and Bethesda Naval [Hospitals] and all across our land—young boys without legs, or arms, or genitals, or faces, or hopes."[40] McGovern's amendment failed, 55 to 39.[41] The Senate was not used to being told it reeked.

Nixon wanted to withdraw later than the December 31, 1971, deadline proposed by McGovern, favored by a majority of Americans, and acquiesced to by the North Vietnamese, as Kissinger had reported to him a month before his speech. "They've extended their deadline now," said Kissinger. "They used to say June 30th this year. Now they're saying December 31st, [1971,] and [in] another three months they may say July 1st[, 1972,] and then we're in business."[42]

At this point, the two men's disagreement had narrowed to a question of whether to exit Vietnam shortly before or shortly after Election Day. Nixon favored before; Kissinger, after. The national security adviser put his position diplomatically: "I have always thought that we should go this year to the North Vietnamese and tell them we'll get everybody out in fifteen months in return for a total ceasefire for that period and the prisoners, but frankly, it's six months too early," Kissinger told Nixon on February 23, 1971. (In fifteen months it would be May 1972, and that was six months before the election, thus "too early.") "I thought we should do it after the election," said Kissinger.

"It's all got to be out by the summer of '72," said the president.

"That's right. That's fifteen months from now," said Kissinger.[43] He could be flexible, even with math.

A withdrawal deadline, the president warned viewers, would not only forsake American prisoners in the North but endanger the lives of

American soldiers in the South, because "we will have given enemy commanders the exact information they need to marshal their attacks against our remaining forces at their most vulnerable." Hard to see why they'd do that. The North wanted American troops to leave, to clear the way for it to take over the South. Engaging Americans in combat while they were heading out the door would be less than astute and more than risky. It was more likely that the North Vietnamese would do what President Nixon himself had predicted in another televised address two years earlier: "They would simply wait until our forces had withdrawn and then move in."[44]

Vietnamization

Kissinger was watching the president's speech on television with his deputy, and Nixon's favorite general, Alexander Haig. Kissinger was the first national security adviser to have a corner office in the West Wing. The chandelier, federal-style wooden furniture, and ceiling-to-floor windows and drapes made his office look like one of the capital's many distinguished products of the early 1800s. It was, in fact, a feat of architectural backdating. Before Nixon, the room had been a bullpen for White House reporters, sectioned into cubicles and florescent-lit.

Nothing in the speech would surprise Kissinger, who had worked on it, substance and wording; he had even served with Haldeman as a two-man preview audience for the staged ad lib conclusion. But the president would spend the rest of the evening on the phone conducting a postspeech review (that is, hearing his own praises sung by a succession of soloists), so Kissinger had to be prepared with credibility-enhancing details.

The biggest challenge of this speech, apart from killing a withdrawal deadline, was making the recent ground offensive by the South Vietnamese army in Laos sound like a success. The president was competing with an indelible picture, seen around the world, of South Vietnamese soldiers clinging to the skids of helicopters, hoping to be carried away from battle. It inspired little confidence and considerable foreboding. Nixon warmed up his audience by recalling the American ground offensive in Cambodia a year earlier: "Let me review now two decisions I have made which have contributed to the achievements of our goals in Vietnam that you have seen on this chart. The first was the destruction of enemy bases in Cambodia."[1] Nixon had sent American troops into Cambodia to put out a fire he had started by accident. Shortly after he took office, the president secretly decided to start bombing along

Cambodia's border with Vietnam. Hanoi used the border areas of both Laos and Cambodia to infiltrate supplies and soldiers into South Vietnam. This was the Ho Chi Minh Trail, named for the Communist revolutionary who had overthrown the French colonial government that had turned the nations of Vietnam, Laos, and Cambodia into a dominion it called French Indochina. The secret bombing of Cambodia, like all attempts to disrupt the Ho Chi Minh Trail, slowed the North Vietnamese down but didn't stop them. Worse, the secret bombing touched off a series of catastrophes that, one year later, threatened to speed the day when the North Vietnamese government in Hanoi would overthrow the South Vietnamese government in Saigon. In short, to avoid getting hit by American B-52s, the North Vietnamese moved deeper into Cambodia, which led to clashes with Cambodian villagers, which destabilized Cambodia's neutralist government, which precipitated a right-wing coup in the capital of Phnom Penh, which posed a threat to Hanoi's infiltration routes, which prompted Hanoi to send troops even deeper into Cambodia, which threatened the new rightist Cambodian government with overthrow, which raised the possibility that North Vietnam would install a pro-Hanoi regime in Phnom Penh, which would have let it turn the Ho Chi Minh Trail into a superhighway, and which just might have made it possible for Hanoi to topple Saigon sooner instead of later. This was the mess Nixon ordered American troops into Cambodia to clean up. (The bombing of Cambodia and the damage it did remained top secret; it was one reason that Nixon dreaded leaks.)

"You will recall that at the time of that decision," Nixon continued, "many expressed fears that we had widened the war, that our casualties would increase, that our troop withdrawal program would be delayed. Now, I don't question the sincerity of those who expressed these fears. But we can see now they were wrong. American troops were out of Cambodia in 60 days, just as I pledged they would be. American casualties did not rise; they were cut in half."[2] American casualties had spiked during the invasion of Cambodia but fell afterward as a result of ongoing American troop withdrawals and the shifting of combat operations to the South Vietnamese.

"Now let me turn to the Laotian operation." It was clever of Nixon to talk about Cambodia before Laos. Public opinion had been divided on the Cambodian invasion at first, but polls taken a few months after showed that, in retrospect, most voters considered it a success.[3] Maybe they would change their minds about the Laotian one as well. "As you know," Nixon said, "this was undertaken by South Vietnamese ground forces with American air support against North Vietnamese troops which had been using Laotian territory for 6 years to attack American forces and allied forces in South Vietnam." After Cambodia, Congress prohibited the president from using American troops for ground offensives on the Ho Chi Minh Trail. American bombers and fighter jets could blast the trail, American helicopters could carry South Vietnamese soldiers into and out of Laos and Cambodia, but no American boots could touch the ground. "Since the completion of that operation, there has been a great deal of understandable speculation—just as there was after Cambodia—whether or not it was a success or a failure, a victory or a defeat. But, as in Cambodia, what is important is not the instant analysis of the moment, but what happens in the future.

"Did the Laotian operation contribute to the goals we sought? I have just completed my assessment of that operation and here are my conclusions:

"First, the South Vietnamese demonstrated that without American advisers they could fight effectively against the very best troops North Vietnam could put in the field."

The question wasn't so much whether the South could fight as whether they would. Early in the offensive, Nixon seemed to have hope. "The main thing I'm interested in is just to be sure the South Vietnamese fight well," he told Kissinger on February 18, 1971. "They're going to be battling here for years to come. I guess if they fight well, North Vietnam can never beat South Vietnam. Never. Because South Vietnam has more people and more—"

"And more equipment," said Kissinger. Neither of them knew then that on February 12 South Vietnamese president Nguyen Van Thieu had ordered his army to stop the ground offensive once his casualties reached three thousand. It had only begun on February 8. Kissinger and Haig already had doubts about the accuracy of reports from the field. The national security adviser was waiting for the right moment to suggest that the president send Haig to Vietnam to check.[4] It came the following week, after Kissinger learned that sensors had detected North Vietnamese trucks moving down a road he'd been told the South Vietnamese had already cut. Haig "could give you a fair assessment of what the hell is really going on," said Kissinger.[5] Nixon agreed.

"So we are achieving an objective that isn't exactly the one we started out with, but I think it will be important," Kissinger told Nixon on February 27. At least they were bleeding Hanoi's supply lines. If "they hold on into April, then the North Vietnamese are in bad shape," said Kissinger.[6] But the South withdrew from Laos before the end of March.

"However Laos comes out, we have got to claim that it was a success," Nixon told Haldeman on March 9. "Those goddamn leaders of theirs, so they get the hell kicked out of them and have to get out—claim a victory. Armies always do that. They claim victories when they lose."[7]

As did he in his April 7 speech: "Second, the South Vietnamese suffered heavy casualties, but by every conservative estimate the casualties suffered by the enemy were far heavier." The North Vietnamese did suffer higher casualties, most of them inflicted by American artillery fire and airpower—not by the South Vietnamese army. The president didn't mention that.

"Third, and most important, the disruption of enemy supply lines, the consumption of ammunition and arms in the battle, has been even more damaging to the capability of the North Vietnamese to sustain major offensives in South Vietnam than were the operations in Cambodia 10 months ago."[8] It would take Hanoi a few months to replace lost supplies.

"Consequently, tonight I can report that Vietnamization has succeeded."[9] Vietnamization was the name Nixon gave his program of training and equipping South Vietnam to defend itself without American troops. The name was Laird's idea, but the policy fulfilled a campaign promise made by both the Republican and Democratic nominees in 1968.[10] In speech after speech, Nixon stressed that Vietnamization would allow him to bring *all* American troops home; Saigon's troops would become strong enough to replace America's. Here's how he put it in 1969: "We have adopted a plan which we have worked out in cooperation with the South Vietnamese for the complete withdrawal of all U.S. combat ground forces, and their replacement by South Vietnamese forces on an orderly scheduled timetable."[11] And in 1970: "Just as soon as the South Vietnamese are able to defend the country without our assistance, we will be gone."[12]

Before announcing the Vietnamization program, Nixon had asked his top military, diplomatic, and intelligence officials how soon South Vietnam would be able to survive without American troops. The answer was unanimous: not for the foreseeable future. Agreed on this point were the Joint Chiefs of Staff, Nixon's secretaries of defense and state, the US embassy in Saigon, the CIA, and the commander of American armed forces in Vietnam, Gen. Creighton Abrams. There was no dissent: "All agencies agree that RVNAF [Republic of (South) Vietnam Armed Forces] could not, either now or even when fully modernized, handle both the VC [Vietcong, i.e., Communist guerrillas in the South] and a sizable level of NVA [North Vietnamese Army] forces without U.S. combat support in the form of air, helicopters, artillery, logistics and major ground forces."[13] There was considerable disagreement among agencies responding to Nixon's first National Security Study Memorandum (NSSM-1), issued on January 21, 1969, on other aspects of the war, but on the question of whether America could strengthen South Vietnam enough for it to survive on its own without American soldiers, there was no dissent. The consensus was no. Nixon knew this before he first announced that "the primary mission of our troops is to enable the South Vietnamese forces to assume the full responsibility for the security of South Vietnam."

Nixon did not treat Vietnamization as a serious strategy.[14] Instead, he used it to justify the actions he deemed politically expedient. When he wanted to keep American troops in Vietnam as long as he needed them there, he could say that Vietnamization required more time to work. When it was politically safe or advantageous to bring the troops home, he could say Vietnamization had worked. If he had actually believed in Vietnamization, he wouldn't have felt the need to keep American troops in Vietnam long enough to prevent the Communists from winning before Election Day 1972.

The next sentence of the speech gave journalists their headline: "Because of the increased strength of the South Vietnamese, because of the success of the Cambodian operation, because of the achievements of the South Vietnamese operation in Laos, I am announcing an increase in the rate of American withdrawals. Between May 1 and December 1 of this year, 100,000 more American troops will be brought home from South Vietnam." ("The troop announcement itself is the best proof that Laos was successful," Nixon told Kissinger on March 19.)

"This will bring the total number of American troops withdrawn from South Vietnam to 365,000. Now that is over two-thirds of the number who were there when I came into office."[15] This would also leave 184,000 American soldiers in Vietnam, more than enough to prevent a preelection collapse of Saigon, while still giving Nixon the opportunity to announce several more troop withdrawals during the campaign.

"The issue very simply is this: Shall we leave Vietnam in a way that— by our own actions—consciously turns the country over to the Communists? Or shall we leave in a way that gives the South Vietnamese a reasonable chance to survive as a free people? My plan will end American involvement in a way that would provide that chance. And the other plan would end it precipitately and give victory to the Communists."[16]

Nixon was getting out of Vietnam around election time, whether the South could survive or not. "They're going to take some raps, but we've got to get the hell out of there," Nixon had told Kissinger a month before the speech. "That's for sure."

"No question," said Kissinger.

"I'm not going to allow their weakness and their fear of the North Vietnamese to—to—to delay us," said the president.[17] He still hoped for a settlement, but with or without one, American troops would be gone. He saw what was coming, little as he liked to say the words.

President Nixon: The South Vietnamese are not going to be knocked over by the North Vietnamese—not easily.

Kissinger: Not easily. [*speaking over President Nixon*] And that's all we could bring about.

President Nixon: And that's all we can do.[18]

"A Nightmare of Recrimination"

On camera, the president chided anyone who shared the attitude he expressed in private: "I know there are those who honestly believe that I should move to end this war without regard to what happens to South Vietnam. This way would abandon our friends. But even more important, we would abandon ourselves. We would plunge from the anguish of war into a nightmare of recrimination."[1]

He was trying to give the Democrats nightmares of recrimination. Briefing congressional leaders at the White House, he warned them not to set a withdrawal deadline unless they wanted to be blamed for Saigon's fall. This was a serious threat to anyone who thought Vietnamization was a fraud and that South Vietnam wouldn't survive without American troops. Later, on March 26, with Kissinger and other aides, Nixon relived this legislative extortion session, complete with dialogue: "I know the date that we're going to be out of there. It's a reasonable date. It's one that I am convinced is the earliest possible date we can get out without risking a South Vietnamese debacle," Nixon said. "If you, on the other hand, decide that you're going to take over and set arbitrary dates," he said, "then you will have to take the responsibility for an American defeat in Vietnam after all these deaths [and] for the Communization of South Vietnam."

The way Nixon told it, he was the one taking all the risk by insisting that the date be left up to him. If he got out and Saigon fell afterward, Congress could lambaste him. "You can just kick the hell out of me. You can say, 'He was wrong. He continued this war for four more years when we could have bugged—got out four years ago, and we still lost it.'" By saying "four years," he tipped his hand a bit. "'If I were a politician,' I said, 'I'd play that game.'"

On the other hand, congressional leaders could say, "We'll support the commander in chief in his best judgment, because we know that he isn't going to keep an American there any longer than he needs to." That was one way they could play it. "But if you play it the other way, I just want you to clearly understand that if there is any arbitrary date set, then I will have no choice but to put the responsibility on the Democrats in the House and the Senate—on them—for losing everything that we fought for in Vietnam and for bringing on a Communist victory."

It's not as if Nixon said that Congress was a nice little branch of government and he'd hate to see anything bad happen to it. He wasn't that subtle. "I said, 'You think you want to fight it out on [an] end date, we'll beat the hell out of you.' Well, they understood it. Understand? They're with us."[2]

The opposition party, for its part, was not displaying great political courage on the subject. "The Democrats say withdraw by the end of the Congress, for Christ's sake," Nixon told Kissinger on February 24. The Senate Democratic Policy Committee had voted to work for total American withdrawal from Vietnam by the time that the Ninety-Third Congress replaced the Ninety-Second in January 1973—two months *after* the election. "What the hell does that prove?" asked the president. "They know damn well we're going to get out by then."[3]

On camera, the president talked about how he made decisions: "One American dying in combat is one too many. But our goal is no American fighting man dying anyplace in the world. Every decision I have made in the past and every decision I make in the future will have the purpose of achieving that goal."

Before closing, Nixon pushed aside his written text. He looked into the camera's eye: "I am often asked what I would like to accomplish more than anything else while serving as president of the United States. And I always give the same answer: to bring peace—peace abroad, peace at home for America." His tone changed. A little less formal, more conversational. "The reason I am so deeply committed to peace goes far beyond political considerations or my concern about my place in history, or the other reasons that political scientists usually say are the motivations of presidents.

"Every time I talk to a brave wife of an American POW"—the camera slowly zoomed out; viewers could see that the president's hands were folded, his text off to the side—"every time I write a letter to the mother of a boy who has been killed in Vietnam, I become more deeply committed to end this war, and to end it in a way that we can build a lasting peace."

Now he spoke more slowly, more thoughtfully. "I think the hardest thing that a President has to do is to present posthumously the nation's highest honor, the Medal of Honor, to mothers or fathers or widows of men who have lost their lives, but in the process have saved the lives of others.

"We had an award ceremony in the East Room of the White House just a few weeks ago. And at that ceremony I remember one of the recipients, Mrs. Karl Taylor, from Pennsylvania. Her husband was a marine sergeant, Sergeant Karl Taylor. He charged an enemy machine-gun single-handed and knocked it out. He lost his life. But in the process the lives of several wounded marines in the range of that machine-gun were saved." It was an act of self-sacrificing heroism. Who could fail to be moved? Who could help but imagine what Sgt. Taylor's family felt?

"After I presented her the medal, I shook hands with their two children, Karl, Jr.—he was eight years old—and Kevin, who was four. As I was about to move to the next recipient, Kevin suddenly stood at attention and saluted." Nixon paused long enough for everyone to picture the scene clearly: a four-year-old who had lost his father in war saluting the commander in chief. "I found it rather difficult to get my thoughts together for the next presentation.

"My fellow Americans, I want to end this war in a way that is worthy of the sacrifice of Karl Taylor, and I think he would want me to end it in a way that would increase the chances that Kevin and Karl, and all those children like them here and around the world, could grow up in a world where none of them would have to die in war; that would increase the chance for America to have what it has not had in this century—a full generation of peace."[4]

(Viewers did not hear what Nixon told Haldeman and Kissinger after rehearsing the conclusion: "If it doesn't move them, screw them.")[5]

The president walked from his office back to the residence at 9:25 p.m. Kissinger called at 9:26. Too soon. Nixon had to settle into his amply upholstered easy chair in the Lincoln Sitting Room, where he had an ottoman for his feet and a phone within reach and a fireplace he could use year-round (with the air-conditioning cranked in the warmer months). It was the smallest room in the residence and had just one comfortable chair. It was his favorite place in the White House.[6]

"Mr. President?"

"Yeah. Hi, Henry."

"This was the best speech you've delivered since you've been in office."

The president demurred.

"This one was *really* movingly delivered," said Kissinger. "And I don't know whether you saw the commentary afterwards."

Nixon scoffed. He never watched television commentators or the evening news. Every morning the White House staff gave him a written summary of broadcast and print coverage tailored to an audience of him. The president said, "I don't care what the bastards say."

"Well, but this is so amazing," said Kissinger. "No one was flyspecking it." The coverage was favorable. "Everyone is saying 'a strong man sticking to his guns.'"

"I'll tell you one thing," the president said. "This little speech was a work of art." Particularly the conclusion. "It was no act, because no actor could do it."[7]

This was a subtle jab at Ronald Reagan. The former movie and TV star spooked Nixon. Less than two years after California elected Reagan governor in 1966—his first run for public office—Reagan launched a last-minute challenge to Nixon for the Republican presidential nomination. By then, Nixon had his delegates sewn up, but Reagan exercised an uncanny pull on the hearts of conservatives, and they were the heart of the new GOP. Reagan had been on the president's mind that afternoon when he and Kissinger were pondering what might happen if their plans for Vietnam didn't work. Kissinger said America might replace Nixon with Gov. George Wallace, D-Alabama, the ardent segregationist turned all-purpose right-wing demagogue and third-party presidential candidate in 1968.

Kissinger: The only consolation we have—but it won't do us any good—is that the people who put us into this position are going to be destroyed by the right.

President Nixon: Damn right.

Kissinger: They're going to be destroyed. The liberals and radicals are going to be killed. This is a basically right-wing country.

President Nixon: I think it is.

Kissinger: You'd get a Wallace—

President Nixon: You'd probably get a Reagan or Wallace, couldn't you? You'd get a Reagan.

Kissinger: I know, but a Wallace without a Southern accent. Or a Reagan with some [unclear] and education. All of our right-wingers have had some defect, but if we could get—if we get someone—

President Nixon: Reagan has enough; he has enough. [*Kissinger attempts to interject.*] They could go for Reagan. He's desperately interested in it. And, too, with a Reagan in here, you could damn well almost get yourself into a nuclear war and get killed. He's the kind with no—he has no judgment.

Kissinger: No judgment.

President Nixon: No—no finesse. No subtlety. It's all—everything is simple. Thank God it's simple on our side at the moment.[8]

Other politicians who were threatened by Reagan's rise also chose to view him as a harbinger of nuclear Armageddon. This allowed them to see themselves as saving the world.

"No actor in Hollywood could have done that [conclusion] that well," said Nixon.

"Mr. President, I had, after all, heard it before," Kissinger said. Nevertheless, "I had a lump in my throat." Haig, too, was "absolutely moved and overwhelmed." The television commentary was the most favorable Kissinger had ever heard.

Nixon called congressional leaders "a miserable lot." He had briefed them at the White House right before the speech. "After you left I stuck it to them," said the president. "I said, 'If Congress wants to take over, that's fine, but then they take the responsibility for this going down the drain, and that is clear, gentlemen.'"

"Right," said Kissinger. "You are saving this country."

If the speech didn't work, Nixon said, "I'm going to find out soon, and then I'm going to turn right so goddamn hard it'll make your head spin. We'll bomb those bastards right out of the—off the earth." (Nixon often talked about bombing North Vietnam. He had an excellent, but secret, reason to think it would solve his political problems.) "I think you agree, don't you?" Nixon asked.

"I think, Mr. President, we have to make fundamental decisions," said Kissinger noncommittally. It was time to end the call. "Congratulations, Mr. President."[9] Not since Marilyn Monroe serenaded JFK in Madison Square Garden had anyone managed to pack as much awe into the words "Mr. President" as Kissinger did every day.

"A Hell of a Shift"

The speech worked perfectly. The big headlines were exactly what the White House wanted: "Nixon Promises Pullout of 100,000 More G.I.'s by December; Pledges to End U.S. Role in War"—*New York Times;* "Nixon Sets 100,000 Troop Cut"—*Washington Post;* "Vietnamization Success—Nixon"—*Los Angeles Times.*[1] The sidebars were even better: "President Recalls Son of a Hero in Ending Speech"—*New York Times;* "Kevin Taylor, 4, Saluted Nixon February and Didn't Forget"—*Washington Post.*[2] The four-year-old appeared on the ABC and CBS evening news.[3] "He'll always remember the president," said Shirley Taylor, Kevin's mother, Karl's widow. "He tells everybody he went to the White House and saw President Nixon because his daddy was very brave."[4]

"Did you see what she said publicly?" Kissinger asked the president. "She was terrific. You'll see it in the News Summary that she said—"

"I don't read the News Summary anymore."

In that case, Nixon had missed one of the biggest stories of his presidency. "I don't know whether you've noticed, incidentally, the Chinese have invited an American ping-pong team—"

Nixon gasped.

"—to visit China. By itself, it doesn't mean a damn thing. On the other hand . . ." Kissinger let Nixon ponder what the invitation could mean for his two-year attempt to start diplomatic relations with the People's Republic of China.

"A lot."

"Exactly," said Kissinger. The papers called it ping-pong diplomacy. It promised a glimpse into a nation America had cut itself off from for two decades. It suggested that Nixon might have the opportunity to make history.

"Well, anyway, getting back to this thing," said Nixon. He wanted to talk more about his speech. "I don't know where the hell Haldeman is."

The old PR hand inside the White House chief of staff was ecstatic. "The Mrs. Taylor story is a—is a dream story," he told the president. Even Dan Rather, the tough CBS White House correspondent, drew raves in the Oval Office, thanks to a network graphic illustrating Vietnamization with a shrinking soldier. The graphic started out big when Rather opened with how many Americans were in Vietnam at the time Nixon took office, then slowly shrank while Rather discussed Vietnamization, so by the time he got to the impact of Nixon's bringing another 100,000 home, Haldeman said, "they had a little, teeny soldier."[5]

Nixon checked with White House Press Secretary Ron Ziegler to see how the Kevin Taylor story was playing. "You had no problem with the press not believing that that was extemporaneous, did you?" the president asked.

"None whatsoever," said Ziegler.

All the stories described the ending as "off-the-cuff" or "extemporaneous," Haldeman said. (He knew it had taken Nixon eight hours to write.)[6]

"I just made some notes on a yellow pad," said the president.

"Right," said Ziegler.

"That's all. That's all," said Nixon. "I really didn't make the decision until late that afternoon, finally, to use it, because I had already given the text out. I decided at that point that I would close without notes."[7] (Nixon had set the staff to work two months earlier on identifying the little boy who had saluted him. It took a while to pinpoint the correct child. "No, no, no, no, no, no, no, it was not a black kid," Nixon had said when he saw the picture of the staff's first guess. "Goddamn it.")[8]

After Ziegler left, Haldeman gave the president his overnight rating—above average for his time slot. "The normal for the time period is 54.9," said Haldeman. "You got a 55.6."[9]

"Nothing matters, Bob," said the president, "unless they're moved."

"You made it hard as hell, after that, for somebody to tell anybody who saw that speech that Nixon's lying to you, or that Nixon doesn't care about the war, or that Nixon's expanding the war," said Haldeman.

"Or that he doesn't care about people dying," Nixon said. "He's a ruthless barbarian."

"Yeah. A killer," said Haldeman. "He's out there killing our boys because he loves to or something."

"It ain't gonna wash," said the president.[10]

Poll results backed them up on this. Nixon's speech flipped public opinion on his handling of the war.

Before the speech: 42 percent in favor, 46 percent opposed.

After the speech: 48 percent in favor, 40 percent opposed.

It was, Haldeman said, "a hell of a shift."[11]

The second-day headlines were not as good. Nixon had made his intentions a little too obvious at the briefing for congressional leaders, and the results were on the front pages of the *Times* and *Post*. Sen. Robert Griffin, R-Michigan, gave away the game: "In a practical sense, he did set a date for ending U.S. involvement: Election Day 1972." Griffin wasn't just any Republican senator. He was the minority whip, the third-highest-ranking Republican in the Senate. Worse for the White House, the top-ranking Republican seemed to back him up. Senate Minority Leader Hugh Scott of Pennsylvania actually quoted Nixon as saying, "I have a date in mind. I have a definite plan and timetable for terminating all U.S. participation in Vietnam." If Nixon did not get out by the end of 1972, Scott said, "another man may be standing on the platform" on Inauguration Day 1973.[12]

"Scott, of course, is so totally irresponsible," Nixon grumbled in the Oval Office. "He's not supposed to say anything, you know? They're not supposed to go out and say a goddamn word."

Haldeman didn't see much of a problem. "Here's what Bobby Griffin says"—the chief of staff referred to the newspaper—"he says, for practical purposes, it's November 1972."

The president *hmm*ed. "Well, that's bad."

They brought in Ziegler and tried to figure out a way to fix it. "Don't escalate this thing. Don't give them any more," said Nixon. "You reject this bullshit to the effect that I've told senators something I wouldn't tell the American people." Ziegler should just stick to the president's speech. "I didn't tell them about a damn thing I didn't tell the American people."

"That's right," said Ziegler.

"You were there through the whole thing," said the president. (In fact, he'd left early.)

"Right," said Ziegler.

The president had a range of dates in mind. "We've got dates everywhere from July [1972] to August to September to October to November to December to January of 1973," Nixon said. "Don't say that."[13] It would not help to have his press secretary proclaim that he was perfectly willing to bring the troops home anytime from a few months before the election to a couple of months afterward.

How to Kill a Withdrawal Deadline

In its quest to kill any bill that would force the president to bring the troops home by the end of 1971, the Nixon White House made clever use of public opinion polls.

After the president's April 7, 1971, speech, White House pollsters asked: Would you favor congressional passage of a bill requiring the president to withdraw all American troops by the end of 1971? The results: 62 percent said yes, 27 percent no. This was exactly the response Nixon did not want.

So the pollsters changed the question: Would you favor withdrawal by the end of 1971 *even if it meant defeat in Vietnam?* When they put it that way, only 39 percent said yes. When pollsters included "defeat" in the question, the December 31, 1971, withdrawal deadline went from having the support of a large majority to being a minority view. This nuance often got lost in debate over the war. A majority wanted to end the war, but at the same time, a majority opposed *losing* the war. Nixon's reelection prospects rested on his ability to convince voters that he was ending the war without losing it.

The pollsters tried asking the question a third way: Would you favor withdrawal by the end of 1971 *even if it meant a Communist takeover of South Vietnam?* Only 27 percent answered yes to this question.

"That's the word," Nixon said, as Haldeman gave him the results on April 14. "We say, 'Communist takeover.'" There was no real difference between "defeat in Vietnam" and "a Communist takeover," but the change in wording made a difference in the response. "They think, 'Oh, Christ, it's lost anyway. You know? You can't win. Get out. Cut your losses.' But 'Communist takeover,' that sort of hits people," Nixon told Haldeman.

They saved the killer question for last: Would you favor withdrawal by the end of 1971 *even if it threatened the lives or safety of the POWs?* Only 12 percent said yes.

Add up all the people who opposed a December 31, 1971, withdrawal deadline if it meant defeat or a Communist takeover or harm to the POWs, and "you've got 71 percent who do not favor bringing all troops home by the end of the year," Haldeman told the president.[1] The polls provided Nixon with a road map to reverse public opinion. They gave him two hot buttons to push, one marked "POWs," the other "Communist takeover."

One can only imagine the results if pollsters had asked about Nixon's real strategy. Would you favor delaying the withdrawal of American troops from Vietnam long enough to ensure that Saigon doesn't fall before Election Day? Even if the Communists would still take over South Vietnam? Even if it delayed the release of American POWs, thereby increasing the risks to their lives and safety? Even if it was certain to result in more American casualties? One can only imagine the response, because Nixon's pollsters didn't ask such questions. Neither did Gallup or Harris. Neither, oddly enough, did the Democratic pollsters.

Two days after his poll briefing, the president appeared in black tie and tuxedo in the Regency Ballroom of Washington's Shoreham Hotel for the annual meeting of the American Society of Newspaper Editors. A panel of six editors asked nice questions.

"Mr. President, I suppose that like most of us, you must have times when you wake up at 3 a.m. or 4 a.m. and you lie there in the predawn darkness and you think," said William Dickinson of the *Philadelphia Bulletin*. "I wonder, sir, if you would tell us what thoughts or worries at this stage of your Presidency come to mind at a time like that?"

Nixon wanted to say he thought about going to the bathroom.[2]

"If there is any one subject that more often than not comes across my mind," Nixon responded, "it is what can I do the next day that will contribute toward the goal of a lasting peace for America and for the world."

"I realize that you do not wish to state a date at this time at which we will withdraw, and that there was some confusion in the press about what Senator Scott said following a meeting at the White House," said Emmett Dedmon of the *Chicago Sun-Times* and *Chicago Daily News*. "Could I ask this question: Is Senator Scott's use of the date January 1, 1973, in your opinion, a practicable goal?"

"Mr. Dedmon, that is a very clever way to get me to answer a question that I won't answer," the president said to laughter. "The date, let me say, cannot and must not be related to an election in the United States." ("I'm not going to say that it's going to be Election Day," he told Haldeman. "I had to nail that.")[3] The president promptly pushed his two poll-tested hot buttons: "I don't want one American to be in Vietnam one day longer than is necessary to achieve the two goals that I have mentioned: the release of our prisoners and the capacity of the South Vietnamese to defend themselves against a Communist takeover." Then he pushed them both a second time: "I have to do what is right for the United States, right for our prisoners, and right for our goal of a South Vietnam with a chance to avoid a Communist takeover, which will contribute to a lasting peace in the Pacific and the world." And a third time: "It will be necessary for us to maintain forces in South Vietnam until two important objectives are achieved: one, the release of the prisoners of war held by North Vietnam in North Vietnam and other parts of Southeast Asia; and two, the ability of the South Vietnamese to develop the capacity to defend themselves against a Communist takeover—not the sure capacity, but at least the chance." And a fourth: "As long as they hold prisoners, and as long as the South Vietnamese have not yet developed the capacity to defend themselves to take over from us the defense of their own country—a capacity that they rapidly are developing—we will have forces there." And a fifth, this time pushing just the hottest button: "As long as they do retain prisoners, no American president could simply remove our forces and remove the threat to them." And a sixth time with the POWs: "We have the responsibility, as long as there is one American being held prisoner by North Vietnam, to have some incentive on our side to get that man released, and that is why we are going to retain that force until we get it, and I think it will work in the end."[4]

Never mind that the president (1) fully expected Hanoi to release the POWs once he withdrew the troops, which (2) he planned to do shortly before or after Election Day 1972, regardless of whether Saigon could withstand a Communist takeover.

A president's public words are news, regardless of whether he means them. "When President Nixon spoke last night in Washington, we got a clearer picture than ever before of how the end of the game may be played in Vietnam," the NBC anchorman John Chancellor told millions of Americans the following day. "He said Americans will be there until he determines that the South Vietnamese have at least the chance of defending themselves. And Mr. Nixon seemed to place perhaps primary importance on the release of the American prisoners."[5] The top, front-page headline of the *Washington Post:* "President Links U.S. Withdrawal to POW Release; Capability of Saigon Also Stressed."[6] Nixon didn't write these stories, but what would he have changed if he could?

"I'm Being Perfectly Cynical"

Nixon promised to end the war one of two ways: by Vietnamization or negotiation. Publicly, he insisted that any agreement with the North guarantee free elections in the South: "We have declared that anything is negotiable except the right of the people of South Vietnam to determine their own future."[1] Nixon's public position and actual position on negotiations differed as much as did his public and actual positions on Vietnamization.

While the United States and North Vietnam held official, publicly announced negotiating sessions in Paris, Kissinger sometimes stole into a working-class neighborhood on the outskirts of the City of Light for secret talks with the Communists. To the national security adviser, the house at 11 rue Darthé where he met with Hanoi's negotiators looked like it belonged to a factory foreman. Small living room, smaller dining room. Two rows of overstuffed chairs faced each other in the living room; there the representatives of Hanoi sat in a line against one wall, the Americans against the other, their toes just a few feet from touching.[2] So far these cramped, clandestine meetings had produced little to nothing, but on May 31, 1971, Kissinger was going to bring them something new.

He called it a ceasefire-in-place, but that term was misleading. During the proposed ceasefire, all American soldiers would leave South Vietnam. The North Vietnamese, however, would remain "in place" right where they were in the South.

When Nixon first took office, he had insisted that Hanoi withdraw its troops from the South in return for his withdrawal of America's. This remained his public position. As recently as his April 7 address, the president had publicly called for "complete withdrawal of all outside forces."³ According to NSSM-1, that's what the South needed to survive. Saigon's forces might be able to handle the Vietcong on their own, but not the North Vietnamese army—not without American ground troops. And Nixon did intend to withdraw *all* American ground troops, through Vietnamization or negotiations. On this point, his public and private positions were the same. He would not maintain a "residual force" of American soldiers in South Vietnam, as President Eisenhower did in South Korea. He couldn't get the POWs released unless he agreed to total American withdrawal; that was Hanoi's price, one that neither he nor any of his advisers ever came up with a way to avoid paying. He couldn't even get a ceasefire from Hanoi as long as he kept American soldiers in South Vietnam. In other words, America would remain at war in Vietnam until he brought the troops home, either through Vietnamization or negotiation.

President Nixon: The idea that we have to keep a residual force in South Vietnam, I'm not really for it. I don't think the South Vietnamese are . . .
Kissinger: Well, it'd be desirable, but I don't—
President Nixon: Face it: I don't think the American people are going to support it and it isn't like Korea somewhere—
Kissinger: [*speaking over President Nixon*] Above all, Mr. President, your reelection is—
President Nixon: Yeah.
Kissinger: —really important for the future.⁴

In the absence of American ground troops, a ceasefire-in-place of the sort Kissinger was proposing would leave the North poised to take over the South. Publicly, Nixon would continue to demand that all outside forces withdraw at the time of a settlement, but in the secret talks, he had dropped the demand for mutual withdrawal in September 1970.⁵ A ceasefire-in-place would go even further. It would give North Vietnamese troops in the South official standing. It would allow them to militarily occupy and govern swaths of South Vietnamese territory.

Two days before the May 1971 round of negotiations, Kissinger and Nixon had an unusually frank discussion in the Oval Office. "It's a unilateral withdrawal," said Kissinger. "We're not asking them to pull out."

Nixon defined what he needed from a settlement. There were three main things: total American withdrawal, release of the POWs, and a ceasefire. "From the political standpoint," the president said, "our major goal is not the ceasefire through '72." Although a ceasefire wasn't paramount, it was still necessary. "That's important, because we don't want South Vietnam to fall," said the president. But another military issue was of greater political importance: "Our major goal is to get our ground forces the hell out of there long before the elections."

"Oh, yeah," said Kissinger.

"And be prepared to bomb the hell out of them in the event that they break it. We can still do that," said Nixon.

Once more, Kissinger warned of the perils of withdrawing too long before Election Day: "The only problem is to prevent the collapse in '72."

"I know. I know," said Nixon. "It won't collapse in '72, though, if we have a ceasefire and the American air force sitting in Thailand and a few other places."

"But don't you think—I have no political judgment—if there were an agreement this summer that said we'd be out with everything by September 1 next year, that this would kill the issue?" asked Kissinger. South Vietnam wouldn't collapse in the two months between September 1 and Election Day.

"September 1 has the ring of—smack dab of politics," said the president. "I'd say August 1 is the latest you could do, because you've got to get it so it doesn't appear as though you're doing it just as the election campaign begins, we're getting out." For political reasons, it had to be a month that didn't look like it was chosen for political reasons.

Kissinger tried July 1, 1972, an exit date that had previously found presidential favor.

"I think that isn't going to make that much difference," said Nixon, particularly with American "air power in Thailand and a few other damn places."

"Yeah, but the Thais won't let us bomb from there anymore after that," said Kissinger.

"Through the ceasefire?"

"No. Maybe through the ceasefire. I doubt it," said Kissinger.

Nixon suggested "a deal that everything but our air power would be gone."

"That's fine."

"In other words, get all ground forces out by July 1 and all air power out by January 1[, 1973]," said Nixon.[6]

All in all, Nixon had three key military demands. Kissinger had "to get them to consider a POW/ceasefire/withdrawal agreement. That's all. Those three things." (In the secret talks, Nixon and Kissinger cleverly insisted on dividing issues into two groups, one labeled "military," the other "political." America and Hanoi should focus on resolving the military questions, they said, and leave the "political" ones for the North and South to work out later among themselves. The question of who would govern South Vietnam was, they said, a political one. In this way they distanced themselves from the ugly fact that it was the central military question of the war as well, since it would be decided by force in the final battle between North and South following Nixon's withdrawal. When others tried this kind of word game, Kissinger liked to say that they had "preempted the categories.")[7]

The president dismissed other parts of their proposal, such as provisions for international supervision of the agreement and an end to North Vietnamese infiltration through Laos and Cambodia. "They're never going to agree to that, Henry. Never."

Kissinger wasn't sure.

"My point is, how much of an effect is the international supervision going to have on them?" asked Nixon.

"Well, just enough so that Cambodia and Laos don't collapse on us during that period."

"Well, I just don't have much confidence in the supervision aspects," said Nixon.

"I don't either," said Kissinger. "It just slows it down a bit." That's all they'd ever been able to do with infiltration. All the bombing and ground offensives had only slowed the movement of soldiers and supplies on the Ho Chi Minh Trail, not stopped them.

President Nixon: So. So.
Kissinger: So we get through '72. I'm being perfectly cynical about this, Mr. President.
President Nixon: Christ, yes [unclear].
Kissinger: If we can, in October of '72, go around the country saying we ended the war and the Democrats wanted to turn it over to the Communists—
President Nixon: That's right.
Kissinger: —then we're in great shape.[8]

After the election, however, Saigon could fall.

Kissinger: If it's got to go to the Communists, it'd be better to have it happen in the first six months of the new term than have it go on and on and on.
President Nixon: Sure.
Kissinger: I'm being very cold-blooded about it.
President Nixon: I know exactly what we're up to.

Kissinger: But—

President Nixon: Here we are. We've got—we're going to have a whole hell—

Kissinger: But on the other hand, if Cambodia, Laos and Vietnam go down the drain in September '72, then they'll say you went into these, you say—you spoiled so many lives, just to wind up where you could've been in the first year.

President Nixon: Yeah.⁹

"We Want a Decent Interval"

Merely delaying South Vietnam's fall until after the election would not solve all of Nixon's political problems. If the South fell in the first months of Nixon's second term, it would still be obvious to voters that he had lost the war. It would have been too late for them to hold him accountable in the voting booth, but he would nevertheless go down in history books as the president who lost Vietnam. His approval ratings would plummet, his second-term agenda would suffer, and his hopes for realigning American politics, for replacing FDR's New Deal coalition with a New Republican Majority, would be damaged. His critics, the ones who said he could not win the war, would be proven right. To avoid personal and political calamity, Nixon needed the South to survive a year or two after he brought the last American troops home. If it lasted eighteen months or so, Saigon's fall might not look like it was Nixon's fault. Kissinger had a special name for this face-saving period of time.

"We want a decent interval," Kissinger scribbled in the margins of the massive briefing book for his July 9, 1971, secret trip to China. "You have our assurance."[1]

Nixon and Kissinger needed the help of Moscow and Beijing to get a "decent interval." Russia and China were the first and second biggest suppliers of military aid to North Vietnam. Hanoi had to listen to them; it didn't have to *obey* them, but it had to listen to them. Through the Russians and Chinese, Nixon and Kissinger could offer Hanoi something valuable in return for a "decent interval"—a clear shot at taking over the South without fear of American intervention. Kissinger planned to raise the issue subtly in his historic first, secret meeting with the Chinese.

Just before entering the airspace of the People's Republic of China, Winston Lord stood up and walked to the front of the plane so he could, technically, be the first American official in China since 1949.[2] The thirty-three-year-old NSC aide's self-conscious history making was nothing compared to his superiors'. The code name for Kissinger's trip, Polo I, invoked Marco Polo's Chinese travels. Polo I was meant to prepare the way for a thoroughly public trip by the president. This meant that Kissinger would see China, meet Zhou Enlai, make history *first*. Nixon had toyed with Kissinger when the invitation finally arrived, tossing out names of other men who could lead the secret trip, like New York governor Nelson Rockefeller.

"Well, he wouldn't be disciplined enough," said Kissinger.

UN Ambassador George H. W. Bush, perhaps.

"Absolutely not," said Kissinger, "he is too soft and not sophisticated enough."

"I thought of that myself," said the president.

"Bush would be too weak."

"I thought so, too, but I was trying to think of someone with a title," said Nixon.

Kissinger had a title: assistant to the president for national security affairs.

"I'd send Haig," said the president.

"Yeah," said Kissinger. "That's what I think." The *deputy* assistant to the president for national security affairs.

"Somebody like that," said Nixon. "I mean, real tough." One more twist of the knife. "It wouldn't have happened if you hadn't stuck to your guns through this period," the president said.

"Well, Mr. President, you made it possible."[3]

The diplomatic opening to China was Nixon's legitimate brainchild. The invitation that came in the spring of 1971 grew from years of careful cultivation, starting before his presidency. In 1967 Nixon wrote in *Foreign Affairs* that "we simply cannot afford to leave China forever outside the family of nations, there to nurture its fantasies, cherish its hates and threaten its neighbors."[4] Once in office, he'd seen the Soviet Union increase the number of divisions it kept along its Chinese border from twenty-four to thirty or more.[5] The hostility building up between the Soviets and Chinese, the two great Communist powers, presented an opportunity for him. Polo I would give birth to triangular diplomacy. Nixon and Kissinger would play China against Russia, Russia against China, and both against North Vietnam.

As big as the geopolitical opportunity was the domestic political challenge. "People are against Communist China, period," Nixon told one White House aide as ping-pong diplomacy took off. "They're against Communists, period. So, this doesn't help us with folks at all."[6]

Americans had haunting memories of the last war. In 1950, Chinese premier Zhou Enlai (the man Kissinger was flying to China to meet) had publicly warned that he would indeed intervene militarily if UN forces under Gen. Douglas MacArthur's command crossed the thirty-eighth parallel dividing North and South Korea. MacArthur had assured President Harry S. Truman that the Chinese would not carry out the threat—and would get "slaughtered" if they did. MacArthur crossed the line in October; in November, Zhou sent Chinese troops across the border in such great numbers that they quickly drove MacArthur's forces back down across the thirty-eighth.[7] China had an overwhelming numerical advantage and refused to tolerate the overthrow of a Communist government on its border—North Korea's in the 1950s and North Vietnam's in the 1960s and 1970s. This is why no American leader suggested sending the US Army across the seventeenth parallel to overthrow the government in Hanoi. No one wanted American soldiers to be buried in a human avalanche. It did not reflect the lack of a will to win, just the simple wish to avoid disaster on a spectacular scale.[8]

The United States employed diversionary tactics to make the North Vietnamese *think* their homeland was under threat during the Laotian offensive, the aim being to scare Hanoi into keeping troops at home in defensive positions so it wouldn't send them down the Ho Chi Minh Trail as reinforcements. The diversion worked well enough to prompt Hanoi to warn that "China will not remain with its arms folded and watch her neighbors being attacked by the United States."[9] The warning led Nixon and Kissinger to discuss the possibility of Chinese intervention on February 18, 1971. "If we went North, if we landed in Haiphong, or if we landed in Vinh or someplace like that, then it's conceivable," said Kissinger. "But I don't think under present circumstances they'd come in."[10] Nevertheless, China looked askance when Nixon moved into its other neighbors, suspending covert contacts with the White House in response to his two ground offensives on the Ho Chi Minh Trail, the 1970 one with American troops in Cambodia and the 1971 one with South Vietnamese troops (and American air support) in Laos.[11]

While Nixon didn't see a lot of votes in the opening to China, it yielded other political benefits. Ping-pong diplomacy could knock the war off the front page. "For every reason, we've got to have a diversion from Vietnam in this country for a while," said Kissinger.

"That's the point, isn't it? Yeah," said Nixon.

"And we need it for our game with the Soviets," said Kissinger.[12] They wanted two summits, with the second one in Moscow to feature the signing of a treaty limiting nuclear weapons—another historic first, if they could get it. The Soviets weren't moving fast enough toward a deal in

Kissinger's secret negotiations with Ambassador Anatoly Dobrynin. The opening to China would give them a push.

It also confounded the Democrats. As Nixon was the first to point out, only Nixon could go to China. He made sure Kissinger told reporters. "Incidentally, I hope you've mentioned the fact no Democrat could have done it," said Nixon.

"Oh, yes," said Kissinger.

"I have done it because, frankly, the hawks trust me," said Nixon.[13]

Once Nixon stopped needling Kissinger about the secret trip to China, he began lecturing him. If the national security adviser got to make first contact with Mao's government, he would be going fully armed with presidential advice. Nixon revealed the secret of his own success in handling the Communists. Two words: cold steel. "I don't fart around," said the president. Kissinger needed to convince the Chinese "that we're just tough as hell."[14] This was a frequent presidential theme.

President Nixon: I'm probably the toughest guy that's been in this office since—probably since Theodore Roosevelt.
Kissinger: No question.[15]

In particular, Nixon wanted Kissinger to stress how tough he could be on Vietnam, to say he couldn't visit Beijing if Vietnam was boiling over. In that case, he might choose to escalate the war.

President Nixon: I want you to put in that this is the man that did Cambodia, this is the man that did Laos, this is the man who will be—who will look to our interests and who will protect our interests without regard to political considerations.
Kissinger: Exactly.
President Nixon: Without regard to political considerations.[16]

But if he did escalate the war in Vietnam, he wanted Kissinger to make sure the Chinese knew it had nothing to do with them. Just as Laos had nothing to do with them. And neither did Cambodia.

Cold steel.

The national security adviser also needed to show some flexibility.

President Nixon: Don't have it so that you've got to go by the book. If he moves in one direction, move quickly to another. In other words, be in the position to move very flexibly with him. So that you don't say, "Well, now, we've got to take this, this, this, and this, and come back to this and this." Keeping them off balance, hitting them with surprise, is a terribly important thing. So be very flexible and don't worry about whether we've cleared it here. If you think there's something outlandish or something to suggest, throw it in. The Communist thinks in very, very orderly terms and very predictable terms, as distinguished from the present American revolutionaries who don't think at all. You know, they just go out and say, you know, four-letter words."[17]

Nixon instructed Kissinger to reveal select bits of American intelligence to the Chinese for effect. "Put in fear with regard to the Soviet [Union]," said the president. "We have noted that our intelligence shows that the Soviet [Union] has more divisions lined up against China than they have against Europe."

"Well, they're undoubtedly going to tape what I say, and I didn't want them to play that to the Soviet ambassador," said Kissinger.

This argument worked on the man who was secretly taping Kissinger at that very moment. "Well, I'd just put it in that there are reports in the press," said Nixon. "Put it that way."[18]

Meeting Zhou

Time stood still for Henry Kissinger as the plane passed over the Himalayas on July 9, 1971. That's how he recalled it ever after, and who would doubt him in that regard? No matter what happened to him after this day, he was now a major figure in American history and in the history of the world. A big limousine awaited him at an airport on the edge of Beijing. It whisked him through city streets, through Tiananmen Square, and into a walled-off, western part of Beijing—an enormous urban park—before depositing him in a stately Victorian guest house on a lake.[1]

At 4:30 p.m., Zhou Enlai arrived. Kissinger found him fascinating. He filled page after page of his memoirs with Zhou, his "piercing eyes," his "gaunt, expressive face," his "immaculately tailored" clothes, his "controlled tension, steely discipline, and self-control, as if he were a coiled spring." The Chinese premier was "one of the two or three most impressive men I have ever met. Urbane, infinitely patient, extraordinarily intelligent, subtle, he moved through our discussions with an easy grace that penetrated to the essence of our new relationship."[2]

Zhou did not waste time. As the two of them sat facing each other, Kissinger began reading a prepared statement from his briefing book: "Many visitors have come to this beautiful, and to us, mysterious land." The premier put up one hand. China would not seem so mysterious, he said, once Kissinger got to know it. At that, the American stopped reading. He never looked at his briefing book again.³

When the conversation turned to Vietnam, Kissinger began dropping hints that a Communist victory would not be unacceptable: "First, we should have a ceasefire for all of Indochina in good faith. Secondly, there should then be a reasonable effort by all the forces in Indochina which exist to settle their differences among each other. Thirdly, we are not children, and history will not stop on the day a peace agreement is signed. If local forces develop again, and are not helped from forces outside, we are not likely to again come 10,000 miles. We are not proposing a treaty to stop history." He outlined the specifics of Nixon's proposal: American withdrawal, POWs, a ceasefire-in-place, and "respect for the Geneva Accords" of 1954. "We do not want the war to start again," said Kissinger.⁴ He was diplomatically vague. Kissinger implied that American reintervention would be unlikely, but he didn't rule it out.

Kissinger and Zhou had talked for hours. Zhou suggested dinner. As the Chinese served dishes staggering in their variety and generosity, Kissinger and his host enjoyed light conversation. Zhou's personality, Kissinger would later write, was luminous. Their conversations were like none he had ever had with any other world leader. Longer. Deeper. Something about the Chinese premier's face made Kissinger feel that Zhou understood him, without any need for translation.⁵

Kissinger revealed this much in his memoirs. What he did not tell his readers was how far he went after dinner, when Zhou suggested they continue their conversation.

Kissinger stopped playing coy on Vietnam, casting subtlety aside. "If the agreement breaks down, then it is quite possible that the people in Vietnam will fight it out," he said. "If the government is as unpopular as you seem to think, then the quicker our forces are withdrawn, the quicker it will be overthrown. And if it is overthrown after we withdraw, we will not intervene."

"But you have a prerequisite for that," Zhou said, "that is, a ceasefire throughout Indochina."

"For some period of time," Kissinger said. "We can put on a time limit, say 18 months or some period."⁶

Summaries of the "decent interval" exit strategy don't get clearer than this. North Vietnam could take over the South without fear of American intervention, as long as it waited a year or two after Nixon withdrew the last of the troops.

And if Kissinger's suspicions were correct, the Chinese had all this on tape. (The quotes here come from a transcript prepared for President

Nixon by NSC aides who accompanied Kissinger on Polo I. It remained classified for the rest of the twentieth century.)[7]

"Old Friends"

Nixon underplayed the announcement beautifully. No big address from the White House, just a couple of minutes broadcasting live from an NBC Burbank television studio on a set decorated only with a podium. At 7:30 p.m. on July 15, he read an announcement that was simultaneously being made in Beijing. China had invited him to visit in 1972; he had accepted with pleasure.

"In anticipation of the inevitable speculation which will follow this announcement, I want to put our policy in the clearest possible context," said the president. "Our action in seeking a new relationship with the People's Republic of China"—he looked the camera in the eye—"will not be at the expense of our old friends." As he said these last few words, he smiled slightly. He called his trip to China "a journey for peace."[1]

Anchormen expressed slack-jawed astonishment. NBC's John Chancellor called it "one of the most stunning developments in foreign affairs that anyone in my generation can remember." Headlines were big and plentiful: "President Will Seek to Break Down 21-Year-Old Great Wall of Hostility"—*New York Times*; "Profound Impact on Foreign Affairs Expected"—*Los Angeles Times*. Democrats and Republicans alike applauded the initiative. NBC contacted the Beijing government and offered to build a television ground station to allow live broadcasts of Nixon in China via satellite.[2]

All public discussion about "old friends" centered on Taiwan. Improved relations between the United States and the People's Republic jeopardized Taiwan's relationship with America and its seat in the United Nations. Vietnam didn't come up much in stories about Kissinger's secret trip, except in a hopeful way: "Ford Sees Indochina Peace Talks Resulting from Visit."[3]

The acclaim was as loud, widespread, and excited as Nixon anticipated, but he had to share it with Kissinger: "Groundwork Laid by Kissinger, Chou in Secret Meeting"—*Washington Post*; "Kissinger Visit Capped 2-Year Effort"—*New York Times*; "Kissinger Sure to Be a Legend: Already a Swinger"—*Chicago Tribune*. A new celebrity cult of Kissinger mushroomed up in the press coverage of foreign policy. The Establishment, demoralized and battered by the Left and the Right for failures at home and abroad, discovered in one of its own a new kind of idol, even a sex symbol of a sort. *Time* put a cartoon on its cover showing Nixon proudly standing on the prow of the ship of state . . . with Kissinger steering. A *New York Times Magazine* piece—"The Road to Peking, or, How Does This Kissinger Do It?"—found Kissinger as fascinating as Kissinger found Zhou. The opening anecdote had him "deep in conversation with this astonishingly beautiful girl, and the girl is looking at him transfixed." That was apparently considered noteworthy. "They talk freely about Henry's 'swinging sex life' in the articles about him," wrote Bernard Law Collier, who decided to join rather than beat them, dishing on Kissinger's dates with actresses (including Jill St. John, a bona fide Bond girl). On a serious note, the *Times* piece raised the question of whether Kissinger was a genius. It ran a lot of pictures of Kissinger with world leaders and movie stars. For Collier, however, Kissinger's appeal was not merely sexual and intellectual; it was "his gift for cutting through the fog and the garbage around an issue," "his incisive, uncluttered honesty."[4]

"He Deserves Our Confidence"

Reagan had no worries about Nixon in China. He told reporters he was happy about the trip and hoped it might be tied to an end of the Vietnam War.[1] Other conservatives in his state condemned the trip. The head of the United Republicans of California, for example, called it "obscene."

If a President Humphrey had made the announcement, "we would rise up in a storm of opposition," Reagan told the fall meeting of the state Central Republican Committee. "Of course we would." He then delivered a near-perfect distillation of a quarter century of Cold War campaign attacks on Democratic presidents. He blamed Franklin Roosevelt and Harry Truman for Josef Stalin's domination of Eastern Europe (with no hint of a suggestion as to how anyone could have prevented it). He blamed Truman for losing Korea, although it was Republican President Eisenhower who settled the war with an armistice in 1953. He blamed JFK for the Bay of Pigs, although Eisenhower and Vice President Nixon shared responsibility for that disgrace. He blamed JFK for failing to achieve victory in the Cuban Missile Crisis, leaving to the listener's imagination what victory in that nuclear confrontation would have involved. Finally, he blamed JFK and LBJ for failing to achieve victory in Vietnam— at which point he noted that Richard Nixon, in contrast, was a Republican. It would have been a *perfect* distillation of Republican Cold War rhetoric if he had blamed Harry Truman for losing China to the Communists, but this was not the time or place for that.

The genius of such rhetoric was in the way it completely sidestepped any responsibility for devising a plan to *save* any of these countries. Republicans realized early on in the Cold War that they could blame the loss of any country on the alleged weakness, cowardice, appeasement, treachery—even treason—of Democrats. The solution would then appear obvious: elect Republicans.

Democrats could try explaining that some problems just didn't have solutions, but voters have never rallied to forthright confessions of impotence. The Democrats tried that approach once, right after the Communists took over China. The State Department issued the China White Paper, a thick volume of declassified documents detailing the history of failed American attempts to shore up the Chinese Nationalists against Mao's revolution. A World War II hero of demonstrated ability, George Marshall, led much of the American effort, first as a general and then as Truman's secretary of state. None of Marshall's credentials as a war hero and patriot stopped Joe McCarthy from accusing him of treason. The dynamic young candidate for Senate from California, Dick Nixon, made a charge that was harder to ridicule but equally untrue, blaming "our State department's policy of appeasing Communists in China." (At the same time, Nixon called on State to "immediately issue a statement that the United States will not recognize Red China"—thereby putting the Truman administration on notice that recognizing China might get it accused of appeasement or worse.)[2]

The White Paper made the case for the hopelessness of the Nationalist cause quite thoroughly. The Democrats may have had more facts on their side, but after the next election Republicans had five more seats in the Senate and twenty-eight more seats in the House on *their* side.[3]

What are facts compared to the glorious vision of victories achievable and disasters avertible by the simple expedient of casting out those accused of weakness and replacing them with their accusers? Put it that way, and the problem with the vision is all too clear. Reagan presented it much more captivatingly:

> We have lived through a period when we saw a Democratic president bring back the bitter fruit of appeasement from Yalta and Potsdam.
>
> We have seen a Democratic president snatch defeat from the jaws of victory in Korea.
>
> We have seen a Democratic president march up to the barricades in the Cuban missile crisis and then lack the will and intelligence to take the last step to victory there.

A Democratic president disgraced us at the Bay of Pigs and Democratic presidents lacked the will and the wisdom to exact a victory for the young Americans who died in Vietnam.

But this is a Republican president who has said only, "I will go and talk, I have no intention of abandoning old friends and allies."[4]

This was a Republican president who had just used the first high-level American contact with China in two decades to explain in detail the circumstances under which he would let the North conquer "old friends and allies" in Saigon. What he said in public was no evidence at all of what he did in secret.

As a sign of Nixon's good character, Reagan pointed to one of his most famous rhetorical achievements: the 1959 "kitchen debate." As Vice President Nixon and Soviet premier Nikita Khrushchev toured a model American kitchen on exhibit in Moscow, they had started trading barbs about their nations' respective ways of life. Reagan applauded "a Republican president who, when he was a vice president, met with another dictator, the dictator of the Russians, in the kitchen in Moscow, listened to his blustering threats against the United States, and then said, 'Try it and we'll kick the hell out of you.'"[5] (Reagan paraphrased liberally. Nixon was not so confrontational in Moscow, saying: "My point was that in today's world it is immaterial which of the two great countries at any particular moment has the advantage. In war, these advantages are illusory."[6] He was making the same point Reagan would make later and better as president: "A nuclear war cannot be won and must never be fought." Before Reagan became president, however, he spent years cornering the market in rhetorical red meat.)[7] "Until there is some showing that Richard Nixon has undergone a massive change of personality, he deserves our confidence, prayers and best wishes."[8]

Nixon's personality had not changed. He was the same man who had won elections by accusing opponents of weakness, appeasement, cowardice, and even treachery.[9] Attacking such vices never stopped anyone from embodying them.

JFK v. Nixon

Democrats could play the "Who lost _____?" game, too. JFK had proved it in 1960. He blamed Fidel Castro's revolution in Cuba on the Eisenhower/Nixon administration—without ever coming up with a way to prevent it (or, as president, to undo it). Arthur Schlesinger, a Kennedy campaign and White House adviser, captured the candidate's moral dilemma and the speed with which it was resolved:

> Cuba, of course, was a highly tempting issue; and as the pace of the campaign quickened, politics began to clash with Kennedy's innate sense of responsibility. Once, discussing Cuba with his staff, he asked them, "All right, but how would we have saved Cuba if we had the power?" Then he paused, looked out the window and said, "What the hell, they never told us how they would have saved China." In that spirit, he began to succumb to temptation.[1]

He succumbed vigorously. The night before his fourth and final televised debate with Nixon, Kennedy blasted the Republican administration for

> doing nothing for six years while the conditions that give rise to communism grew ... ignoring the repeated warning of our ambassadors that the Communists were about to take over Cuba ... standing helplessly by while the Russians established a new satellite only ninety miles from American shores ... two years of inaction since Castro took power ... this incredible history of blunder, inaction, retreat and failure.[2]

What made front-page headlines, however, was Kennedy's declaration: "We must attempt to strengthen the non-Batista democratic anti-Castro forces in exile, and in Cuba itself, who offer eventual hope of overthrowing Castro. Thus far these fighters for freedom have had virtually no support from our government."[3] Debate continues over whether JFK knew at that time that the CIA was training and equipping anti-Communist Cubans to overthrow Castro, but there's no question that calling for such action as a candidate made it more difficult for Kennedy to reject it as president. He was unwittingly laying the political groundwork for the Bay of Pigs debacle.

Kennedy also accused Nixon and Eisenhower of letting America lose its nuclear superiority. Kennedy charged that because of Eisenhower's neglect, the Soviets had, or would soon have, a larger arsenal of long-range nuclear missiles than the United States.[4] The "missile gap" was a bogus issue. Eisenhower denied it (and learned how little good such denials do).[5] Kennedy was in the White House less than a month before his new secretary of defense, Robert McNamara, told reporters that there was no missile gap in the Soviet Union's favor, though there might be one in America's.[6]

In their first televised debate, Kennedy used his opening statement to take the discussion off the agreed-upon topic of domestic policy and mount a hawkish attack on Eisenhower/Nixon foreign policy. Kennedy contended that America's survival as a free nation was at stake:

> In the election of 1860, Abraham Lincoln said the question was whether this nation could exist half-slave or half-free. In the election of 1960, and with the world around us, the question is whether the world will exist half-slave or half-free, whether it will move in the direction of freedom, in the direction of the road that we are taking, or whether it will move in the direction of slavery.... We discuss tonight domestic issues, but I would not want [there] to be any implication to be given that this does not involve directly our struggle with Mr. Khrushchev for survival.... Are we as strong as we must be if we're going to maintain our independence, and if we're going to maintain and hold out the hand of friendship to those who look to us for assistance, to those who look to us for survival? I should make it very clear that I do not think we're doing enough, that I am not satisfied as an American with the progress that we're making.[7]

Nixon counterattacked as best he could. He accused the Democrats of the Truman era of leaving a "dangerous missile gap which they handed over to President Eisenhower to straighten out."[8]

Neither Kennedy nor Nixon, both World War II Navy lieutenants, could count on their military service to protect them from charges of being weak or worse. Gen. George Marshall's role in winning World War II didn't save him from becoming an answer to the question, "Who lost China?" Gen. Dwight Eisenhower's larger role in winning World War II didn't save him from becoming an answer to the question, "Who lost Cuba?" If a general's stars could not fend off such attacks, neither could a lieutenant's bars. Kennedy didn't have a stronger defense than Nixon, but he did manage a stronger rhetorical offense. In 1960, both nominees tried to out-hawk each other. Kennedy won with "Who lost Cuba?" and the mythical missile gap. He didn't win fair and square, but he won.

The defeated candidate repaid the new president in full. If JFK could accuse Eisenhower and Nixon of losing Cuba, then Nixon could accuse JFK of losing an opportunity to overthrow Cuba's Communist government. After the Bay of Pigs, Nixon said Kennedy looked "both weak and aggressive at the same time."[9] Nixon made a standard hawk argument:

if force didn't work, it meant that too little was used. In Cuba, "when the critical moment came, when we had to decide whether we needed additional power to make the project succeed, we decided not to go through with the project we had started," said Nixon.[10] Since he didn't say what kind of additional power would be needed or what America would do if it succeeded in overthrowing Castro (occupy the country? allow a largely pro-Castro populace to vote on a successor?), no one could prove him wrong.

Nixon also accused JFK of losing ground in Laos. During his first one hundred days, Kennedy decided to seek a coalition government for the kingdom of Laos. Coalition governments share power among competing factions. In Laos, that meant sharing it with the Communist insurgency, the Pathet Lao.[11] Once more, Nixon accused Kennedy of weakness, calling the coalition government "a victory for Communist power."[12] (In 1975, Laos would be the last part of Indochina to go Communist, after both Cambodia and South Vietnam.)[13]

Kennedy, however, avoided the accusation that he lost Vietnam. When he sent American helicopter pilots and military advisers to Vietnam in 1962, Nixon criticized other Republicans for criticizing the president:

> I don't agree at all with any partisan or other criticism of the United States build-up in Vietnam. My only question is whether it may be too little or too late. It is essential that the United States commit all the resources of which it is capable to avoid a Communist takeover in Vietnam and the rest of Southeast Asia.
>
> I support President Kennedy to the hilt and I only hope he will step up the build-up and under no circumstances curtail it because of possible criticism.[14]

Within this seeming defense of JFK, Nixon hid a hawkish attack. He subtly cast Kennedy as weak ("I only hope he will step up the build-up") and started placing blame for the buildup's failure before it even occurred ("My only question is whether it may be too little or too late"). The "too little too late" attack is a classic: it assumes that there is a certain amount of force that will achieve victory, blames the attacked for failing to apply it, and sidesteps all responsibility for coming up with an actual strategy. How could JFK respond to that? By saying that he was doing enough? He would disprove that by continuing to increase the number of American advisers in Vietnam, which rose from 600 when he took office to 16,000 at the time of his death. That was enough to keep him from becoming the answer to the question "Who lost Vietnam?" in his first term. It was not enough, however, to accomplish the mission he defined as training South Vietnam's army to defend itself without American soldiers. Nothing was.

How else could JFK have replied? By admitting that no amount of American military involvement would make South Vietnam capable of defending itself? That would have given Nixon and others the opportunity to accuse him of defeatism, weakness, cowardice—the standard litany. It would also have forced JFK to explain why he was sending American training troops on a mission that was futile and, for some, fatal.

JFK did have another alternative. If he had lived and won reelection in 1964, Robert Kennedy told one interviewer, "We would have handled [Vietnam] like Laos." This would have come as a surprise to Kennedy administration officials who had heard JFK rule out a coalition government for South Vietnam.[15] A coalition was, however, the only readily available alternative to war, if only because North Vietnam was willing to settle for one. If, however, a coalition led eventually to a Communist government in the South, a president who had agreed to it would be accused of losing Vietnam.

Robert Kennedy fuzzed his own position on a coalition government during his 1968 presidential campaign. When Sen. Eugene J. McCarthy, D-Minnesota, called on the Johnson administration to state publicly that a coalition government would be acceptable in South Vietnam, RFK said, "I would be opposed to what I understand is Senator McCarthy's position to be of forcing a coalition government on the government of Saigon, a coalition with the Communists, even before we begin the negotiations."[16] RFK sounded like he was against a coalition government at that time, but he didn't rule out the possibility of forcing Saigon to accept a coalition *after* negotiations began.

By June 1971, Ted Kennedy, the last Kennedy brother, was not only casting a coalition government as a path to peace but also suggesting that this was something his brother Robert had favored in his last campaign.

> There will be no peace, there will be no end to killing, until the conclusion, first stated by a man who died three years ago this week—the concept of a coalition government—is met and understood. In those days, that conclusion was ridiculed as placing the fox in the chicken coop. The ridicule was unjust. But today the concept of a coalition government is called "defeat." And that is hypocrisy.[17]

Nixon, however, rejected a coalition government outright. In the final week of his 1968 presidential campaign, he said that imposing a coalition government on Saigon would be nothing more than a "thinly disguised surrender."

> To the Communist side, a coalition government is not an exercise in cooperation, but a sanctuary for subversion. Far from ending the war, it would only insure its resumption under conditions that would guarantee Communist victory.[18]

A "decent interval" deal would have all these flaws as well. It, too, would be imposed on Saigon against its will. It, too, would lead to resumption of the war under conditions (the absence of the American military) that would lead to Communist victory. It, too, would be a disguised surrender.

A "decent interval" had one clear political advantage over a coalition government: the "decent interval" would be a better-disguised surrender. On paper, it would deny the Communists any role in Saigon's government unless and until they won an election. (They would never get a chance to compete at the ballot box, because Saigon would not allow such an election to take place.)

A coalition government, on the other hand, had one clear advantage over a "decent interval": Nixon could have gotten a coalition government in his first year as president. Hanoi started demanding a coalition government in 1969 and continued to do so for the next three years. If Nixon had agreed in 1969, he would have avoided four years of war and saved the lives of 20,000 American soldiers as well as hundreds of thousands of Vietnamese, North and South. He could even have provided safe passage out of the country for the thousands of Vietnamese who fought or worked on the American side as soldiers or civilian government employees. The human cost of a coalition government would have been much lower.

The political cost, however, would have been much higher. If Nixon had accepted a coalition government for South Vietnam in 1969, he would have left himself open to the attacks he had made on JFK for accepting one for Laos in 1961. Nixon's opponents could have quoted him calling a coalition government "a victory for Communist power." If the coalition government led to a Communist one, as Nixon predicted, then his name would have become the answer to "Who lost Vietnam?" He could have saved thousands of lives, but he would have lost his reelection campaign.

Nixon publicly dismissed a coalition government as merely a Communist demand "that we overthrow the Government of South Vietnam as we leave."[19]

The Kennedy Critique

Ted Kennedy did more than issue a call for a coalition government. He actually accused Nixon of timing America's military exit from Vietnam to the president's reelection campaign. Although Kennedy remained the most prominent liberal Democrat in America following the deaths of his brothers—and the charge he made was accurate—he drew a response from other American liberals that illustrates the feckless, floundering disarray in which liberalism remained mired during the early 1970s.

Kennedy made his penetrating critique of the president's strategy in a June 7, 1971, speech to the National Convocation of Lawyers to End the War, the same speech in which he called for a coalition government. "At last, the ultimate and cynical reality of our policy is beginning to dawn on the American people. The only possible excuse for continuing the discredited policy of Vietnamizing the war, now and in the months ahead, seems to be the president's intention to play his last great card for peace at a time closer to November 1972, when the chances will be greater that the action will benefit the coming presidential election campaign," said Kennedy. "How many more American soldiers must die, how many innocent Vietnamese civilians must be killed, so that the final end to the war may be announced in 1972 instead of 1971?"[1]

Kennedy's charge was accurate and his question on point. But the last of the Kennedy brothers had no documents to back his words. (Nixon's White House tapes remained secret and under the president's complete control; the NSC transcripts of Nixon and Kissinger's negotiations with foreign leaders were also top secret.)

Although Kennedy lacked inside information about the Nixon White House, he did have some about how the Kennedy White House had handled the war. At this point, so did much of America. The source was Kenneth O'Donnell, who had worked on JFK's first congressional race, helped run his 1960 presidential campaign, and served as a top White House political aide. In the August 7, 1970, issue of *Life* magazine, O'Donnell quoted a meeting between President Kennedy and Mike Mansfield, the Senate majority leader, in the spring of 1963. The United States needed to withdraw from Vietnam, said Kennedy. "But I can't do it until 1965—after I'm elected." After Mansfield left the room, the president demonstrated that he wasn't just telling Mansfield what the senator wanted to hear by elaborating on his intentions to O'Donnell: "In 1965, I'll be damned everywhere as a Communist appeaser. But I don't care. If I tried to pull out completely now, we would have another Joe McCarthy red scare on our hands, but I can do it after I'm reelected.

So we had better make damned sure that I *am* reelected."² If this was indeed JFK's approach, then Nixon wasn't the first president to time withdrawal from Vietnam to his reelection.³ During the Kennedy administration, 168 Americans died in Vietnam, 100 of them during 1963. The difference between JFK's approach and Nixon's was not one of kind, but of degree.

O'Donnell and Dave Powers, another Kennedy White House aide, said they once asked JFK how he would get out of Vietnam without suffering a loss of American prestige. "Easy," said the president. "Put a government in there that will ask us to leave."⁴ Daniel Ellsberg put the question to Robert Kennedy while the defense analyst was working on the Pentagon Papers: Would JFK have been willing to accept a Communist takeover of South Vietnam in 1965? According to Ellsberg, RFK replied: "We would have fuzzed it up. We would have gotten a government in that asked us out or that would have negotiated with the other side." (This was the same interview in which RFK, according to Ellsberg, said President Kennedy would have handled Vietnam like Laos, where JFK had accepted a coalition government.)⁵

Other Kennedy administration officials remained unconvinced. Former secretary of state Dean Rusk said that the idea that JFK had decided to get out of Vietnam following his reelection was hard to believe "for one *unimportant* reason and for one very *important* reason." The "unimportant reason" was that the president never mentioned any such decision to Rusk. "This by itself is not conclusive, since for reasons of his own, Kennedy possibly didn't want to confide in me his future plans for Vietnam," Rusk wrote in his memoir, *As I Saw It*. The important reason, Rusk said, was that "if he had decided in 1962 or 1963 that he would take the troops out after the election of 1964, sometime during 1965, then that would have been a suggestion that he would leave Americans in uniform in a combat situation for domestic political purposes, and no president can do that."⁶ It would be nice to believe this, but the Nixon tapes prove that at least one president *did* do that—and he did it without confiding in his secretary of state or secretary of defense.

Unsurprisingly, Republicans counterattacked against Ted Kennedy's charge. More than 1,200 at a party fund-raising dinner in Boston cheered as Gov. Ronald Reagan denounced him: "The senator has charged that the president plays politics with the lives of young Americans. This is an irresponsibility that no Republican has ever committed against the five Democratic presidents who presided over the four wars we've known in our lifetime."⁷ (Reagan was a master of Republican rhetoric but a bit blind to the ways of other practitioners of the art. Accusing Democratic presidents of playing politics with American lives was all too common.)⁸ The next day, Reagan called on Kennedy to apologize for making such a "venal, vicious charge."⁹

Stunningly, one of the few people who could compete with Kennedy for the title of America's top liberal rose to Nixon's defense: Hubert Humphrey, Nixon's 1968 opponent and potentially his 1972 one as well. "I do not believe the President is playing politics with Vietnam," Humphrey said on the floor of the Senate.¹⁰ "I do not think it is a matter of the president being cynical or of the president trying to prolong the struggle." Humphrey supported a congressionally set withdrawal deadline, but he saw Nixon's opposition simply as a difference of judgment. "I do not think that makes him bad," said Humphrey. "I may disagree with the president's policy. I do not disagree with his sense of sincerity or integrity."¹¹

The former vice president made the next day's *New York Times:* "Humphrey Says Nixon Shuns Politics on War."¹² He also got a thank-you call from the president.

President Nixon: I just want to say, if you got a little flack from your party, just know there was one party down at the White House that was appreciative.

Humphrey: Well, I very much appreciate that. [*President Nixon chuckles.*] And let me say I did exactly what I would have expected that you or someone like you to have done under the same circumstances, and which you did do.

President Nixon: Yeah.

Humphrey: And I believe there are rules of fair play, and I'm not going to—I have too much respect for the office and the man that occupies it [*President Nixon attempts to interject*] to permit things like that to go unchecked.¹³

The president complimented Humphrey on a recent Gallup poll showing him neck and neck with Kennedy for 1972. As for Vietnam, "there's no political mileage in [sic] anybody in keeping this going," said Nixon.

"Of course not," said Humphrey.

The president laughed.¹⁴

Anyone who wondered why Nixon didn't just withdraw from Vietnam upon taking office had only to look at the speech Humphrey gave on national television as the 1968 Democratic presidential nominee.

> Let me first make clear what I would not do.
> I would not undertake a unilateral withdrawal.

To withdraw would not only jeopardize the independence of South Vietnam and the safety of other Southeast Asian nations. It would make meaningless the sacrifices we have already made.
It would be an open invitation to more violence . . . more aggression . . . more instability.
It would, at this time of tension in Europe, cast doubt on the integrity of our word under treaty and alliance.
Peace would not be served by weakness or withdrawal.[15]

If Nixon had brought all the troops home in 1969, Humphrey could have attacked him from the right, called him weak, accused him of inviting aggression, violence, and instability. Of destroying American credibility. Of snatching defeat from the jaws of victory. Of sounding a retreat that meant 31,000 Americans had died in vain.[16]

Any Democrat could have made that charge, since *none* of the Democratic Party's 1968 presidential candidates had called for immediate withdrawal—not Robert Kennedy, or Eugene McCarthy, or even George McGovern, who entered the race after Kennedy's assassination. That's why the notion that Nixon could simply have withdrawn from Vietnam shortly after taking office without suffering any political consequences is a fantasy. Leaving Vietnam may have been a popular idea, but losing it was not.[17] Since Nixon realized that leaving meant losing, and that losing the war meant losing the next election, he viewed liberal attempts to set a withdrawal deadline as political poison. Unless he was willing to be a one-term president—hardly the worst thing to be—Nixon had to keep Saigon from falling before Election Day 1972.

Nixon portrayed his decision to continue the war as an act of political courage: "There were some who urged that I end the war at once by ordering the immediate withdrawal of all American forces. From a political standpoint this would have been a popular and easy course to follow."[18]

It would have been, in fact, political suicide. In 1969, the Republican National Committee (RNC) commissioned a secret poll to gauge the popularity of various Vietnam exit strategies. Immediate withdrawal came in last place. The statement—"We should get out immediately on North Vietnam's terms"—drew the opposition of 83 percent. By comparison, "We should use atomic bombs to end the war in Vietnam quickly" was opposed by 82 percent. In Nixon's first year in office, immediate withdrawal from Vietnam polled worse than nuclear war.[19]

To be fair to the American people, no one had told them that it was the unanimous view of the Joint Chiefs, CIA, State, Pentagon, and Gen.

Creighton Abrams that South Vietnam's survival would depend on American troops for the foreseeable future, meaning that America's true alternatives in Vietnam came down to either staying indefinitely or leaving and losing. Nixon seemed to offer a third way, Vietnamization, a promise that all the troops would come home once he judged South Vietnam capable of its own self-defense.

"Precipitate withdrawal," he warned, would lead to dire consequences. In South Vietnam, it would mean Communist takeover and the slaughter of thousands; in America, "as we saw the consequences of what we had done, inevitable remorse and divisive recrimination." All this was as true in 1969 as it would be four years later, although the phrase "precipitate withdrawal" implied that at some point leaving would not mean losing. The president called on "the great silent majority of my fellow Americans" to be "united against defeat," knowing full well that it already was.[20] The Silent Majority speech raised his approval ratings to the highest level of his presidency, 68 percent.[21] He would not equal it again for four years, when he finally announced that he had achieved "peace with honor" in Vietnam and the troops and prisoners were coming home.[22]

Until then, Nixon made a political virtue of political necessity: "It is tempting to take the easy political path: to blame this war on previous administrations and to bring all of our men home immediately, regardless of the consequences, even though that would mean defeat for the United States" (April 30, 1970);[23] "From a political standpoint, this would be a very easy choice for me to accept" (May 8, 1972).[24] Immediate withdrawal would have been the easy way for Nixon to become a one-term president.

The Liberal Mistake

Ted Kennedy had shown Democrats a way to have at Nixon—something they desperately needed. According to Kennedy's critique, Nixon's decision to prolong the war wasn't an act of political courage, just political calculation—even political cowardice. Nixon wasn't going to deny victory to the Communists, just delay it until politically convenient. Beneath the appeals to patriotism lurked boundless political opportunism. What could be less patriotic than sacrificing the lives of American soldiers for political gain?

Liberals other than Kennedy, however, were making a different charge against Nixon. For some reason, the most vocal, influential, even brilliant ones grew convinced that Richard Nixon would never let the Communists take over South Vietnam. Ever. Not merely before his reelection, but for as long as he remained president. This was the serious opinion of serious men.

Three of them—Paul Warnke, Morton Halperin, and Leslie Gelb— were, by any measure, true experts on Vietnam. All had worked in the Pentagon during the Johnson administration, first under Defense Secretary Robert McNamara, then his successor, Clark Clifford. They were the top three officials in the Pentagon's Office of International Security Affairs, the in-house think tank known as "the State Department within the Defense Department" that advised the defense secretary on the many international implications of matters military. They were brilliant men with impeccable academic credentials, advanced Ivy League degrees all around: Warnke's from Columbia, Halperin's from Yale, Gelb's from Harvard. When McNamara decided he needed an in-depth study of American decision making in Vietnam, a top secret history of the war destined to become known as the Pentagon Papers, the assignment fell, quite naturally, to Warnke, Halperin, and Gelb. David Halberstam may have won a Pulitzer Prize for *The Best and the Brightest*, his epic study of how brilliant men sucked America into a quagmire, but Warnke, Halperin, and Gelb wrote theirs first.

Gelb was one of Kissinger's grad students at Harvard, Halperin the guest lecturer at the professor's last class. The new national security adviser took Halperin with him to the White House as the Vietnam expert on the National Security Council (NSC) staff. This gave Halperin insider status in two administrations, one Democratic, one Republican (although he never joined the inner circle of either president). The experience enhanced Halperin's credibility as a Vietnam expert, and for many liberals his stature grew after he severed ties with the White House in the spring of 1970 over Nixon's decision to send American troops into Cambodia. By then, Gelb was a Fellow at the Brookings Institution, and Warnke had opened up a law practice with Clifford, the former defense secretary whom all three continued to advise. One result of this advice was an article, "Clark Clifford on Vietnam: Set a Date and Get Out," that *Life* magazine put on its May 22, 1970, cover. Clifford called for complete withdrawal from Vietnam by December 31, 1970.[1] (The White House learned of this in advance through an illegal FBI wiretap on Halperin's phone, placed there on the never-substantiated suspicion that he had leaked the secret bombing of Cambodia while on Kissinger's NSC staff. "This is the kind of early warning we need more of," Nixon's chief domestic policy adviser, John Ehrlichman, wrote Haldeman, who agreed.)[2]

Demand for Warnke's, Halperin's, and Gelb's advice grew among Democrats who wished to replace Nixon in the White House. (Even after Halperin became an adviser to Edmund Muskie, the Democratic front-runner, the FBI tap remained on his phone.)[3] They influenced politicians and the public through their writing as well.

They thought Nixon would withdraw from Vietnam only when and if Congress forced him. "The president has not sought to keep his policy a secret from the American people," Gelb and Halperin wrote in the *Washington Post* on May 24, 1970. "But despite what he said, most observers have assumed that he was first of all a politician, and that he therefore had ruled out any escalation and was planning to withdraw all American forces before 1972." This revealed more about the authors than the president, specifically that (1) they thought it would be politically advantageous for Nixon to bring the last troops home before 1972, even though that would make it possible for the Communists to overthrow the South Vietnamese government before Election Day—a political calamity for an incumbent president; and (2) they did not know of the secret, RNC-commissioned poll indicating that one particular military escalation would be spectacularly popular, and therefore *not* the kind of thing a politically motivated president would "rule out." The authors' convictions were stronger than their evidence: "Neither by word nor deed has the president indicated any intention to withdraw all American forces." In fact, the president *had* indicated it "by word," having spoken of "complete withdrawal of all U.S. combat ground forces," "withdrawal, first, of all U.S. combat troops," and "this government's acceptance of eventual, total withdrawal of American troops" in three separate, nationally televised addresses.[4] Despite Nixon's frequently made promise, the authors concluded that he would withdraw "only to the degree that he believes the South Vietnamese forces can fill the gap, and only if Hanoi does not step up its military effort. Given these criteria, there will be many American troops in Vietnam for a long time to come."[5]

The dilemma was apparent. If Nixon really intended to withdraw only when South Vietnamese troops could do the same job as American ones, then he would have to keep American troops in Vietnam indefinitely. On the other hand, if Nixon really intended to bring all the troops home, then he would have to do so despite South Vietnam's inability to take their place in battle. Nixon could really intend only one of those two things. The liberals just picked the wrong one. They didn't see Nixon as a politician who would withdraw all the troops merely to create the semblance of the "peace with honor" he had promised (knowing

Hanoi would not agree to a ceasefire or release of the POWs for anything less than total American withdrawal). They saw Nixon as more than a politician. They thought he would keep Americans fighting and dying as long as it took to preserve South Vietnam. "If the President means what he says, we will have American forces numbering perhaps 50,000, perhaps 200,000, in Vietnam indefinitely, propping up the current Saigon regime," Gelb and Halperin wrote in the *Post* on October 11, 1970. "The fighting will continue and Americans will continue to die and be wounded."[6]

Warnke took up the same theme in the June 21, 1970, *Post,* referring to "the President's unstated (but now undisguisable) insistence that our proxy regime must be permanently secured."[7] If this were true, the war would never end.

Clifford publicly predicted that if Nixon won reelection, the Vietnam War "will continue indefinitely." The former defense secretary told the Democratic Platform Committee, "The Republican Party, which in 1968 pledged an end to the war, in 1972 offers what appears to be perpetual war."[8]

The President of Perpetual War was a poor choice of political narratives, primarily because it was not true. Perpetual war was a harsh charge, to be sure, but it would backfire at the worst possible moment, when Nixon traded complete American military withdrawal from Vietnam for the POWs before Election Day. Yet it dominated liberal and Democratic rhetoric through the 1972 campaign.

The party's ultimate presidential nominee believed. "Vietnamization with its embrace of the Saigon regime perpetuates the war," George McGovern told anyone who would listen throughout his campaign for the nomination and presidency.[9] "The president's Vietnamization policy virtually guarantees that our prisoners will remain in their cells, that our troops will remain in danger, that the negotiations will be stalled, and the killing will continue."[10]

The Nixon that haunted the liberal imagination was someone who took anti-Communism to an extreme. Someone who, for example, would withdraw American troops from Vietnam only if Saigon's were strong enough to take their place, even after it became clear that this would mean sacrificing American lives in perpetuity.

By casting Nixon as the implacable foe of Communist victory in Vietnam, liberals played into his hands. If he was the man they thought, he would never quit, never give up or give in. But Nixon would do all these things. And when he did, liberal campaign rhetoric would provide him with cover.

One of the brilliant young liberals who grew convinced of Nixon's undying devotion to Saigon was Daniel Ellsberg. His was a résumé to shame an overachiever: Harvard summa cum laude, a year abroad at the University of Cambridge, company commander in the Marine Corps, nuclear strategist, RAND Corporation think tanker, Pentagon analyst, a couple of years in Vietnam working on pacification before returning to the States and being asked to write a chapter of the Pentagon Papers, the top secret Defense Department history of the Vietnam War. Ellsberg is best known as the man who turned the Pentagon Papers over to the newspapers in 1971—the biggest unauthorized disclosure of classified information up to that time.

Less known is why he leaked them. Ellsberg had been hopeful when Nixon chose Kissinger as national security adviser. Ellsberg knew Kissinger as a realist about Saigon's prospects. In *Secrets: A Memoir of Vietnam and the Pentagon Papers*, Ellsberg wrote:

> In 1967 and 1968 I had been with him in conferences on Vietnam, where he was expressing a point of view that was well in advance of that of any other mainstream political figure at that point. He argued that our only objective in Vietnam should be to get some sort of assurance of what he called a "decent interval" between our departure and a Communist takeover, so that we could withdraw without the humiliation of an abrupt, naked collapse of our earlier objectives. He didn't spell out how long such an interval might be; most discussions seemed to assume something between six months and two years.[11]

Ellsberg didn't want a "decent interval." He thought the war should end immediately. But he saw Kissinger's presence on Nixon's White House staff as a sign that at least the administration was getting out of Vietnam. That was before one of Ellsberg's closest friends, the White House insider Morton Halperin, convinced him that Nixon was staying in. Listening to Halperin, Ellsberg grew convinced that Nixon was doing what presidents before him had done. Ellsberg's reading of the Pentagon Papers convinced him that Johnson, Kennedy, Eisenhower, and Truman had all escalated America's involvement in Vietnam to keep it from falling on their watch.

> Nixon had no readiness at all to see Saigon under a Vietcong flag after a "decent interval" of two or three years—or ever. Not, at least, while he was in office. That meant not through 1976, if he could help it, as he believed he could. That didn't mean he expected the VC [Vietcong] or

DRV [North Vietnam] to give up, permanently, its aim of unifying the country under its control. And so it meant that the war would essentially never end. His campaign promise of ending the war was a hoax.[12]

Nixon's exit strategy for Vietnam was indeed a hoax, but not the one that Ellsberg and the liberals feared. Their certainty that Nixon would not settle for a "decent interval" and that he would never let Saigon fall made it easier for him to do both. They unwittingly reinforced Nixon's image as a president who would never retreat, never surrender, a steadfast ally who would never accept a Communist takeover of Vietnam. Their attacks made him stronger.

"Super Secret Agent"

In January 1972, Nixon revealed that secret negotiations had taken place in Paris between the North Vietnamese and Kissinger. After that, Kissinger made the cover of *Time* and *Newsweek*. A front-page headline in the *Times* referred to him as a "Super Secret Agent."[1] And in February 1972, when Nixon finally did go to China, Chairman Mao Zedong said during their first meeting: "We two must not monopolize the whole show. It won't do if we don't let Dr. Kissinger have a say. You have been [made] famous about your trips to China."

"It was the president who set the direction and worked out the plan," Kissinger said.

"He is a very wise assistant," Nixon said, "to say that."

Mao laughed.

"He doesn't look like a secret agent," Nixon said. That could be taken more than one way. The president started praising his adviser, sort of. "He is the only man in captivity who could go to Paris 12 times and Peking once, and no one knew it—except possibly a couple of pretty girls."

"They didn't know it," Kissinger said. "I used it as a cover."

"Anyone who uses pretty girls as a cover must be the greatest diplomat of all time," Nixon said. It was as if he was parodying what the papers were saying about Kissinger.

"So," Mao asked, "you often make use of your girls?"

"*His* girls, not mine," Nixon said. "It would get me into great trouble if I used girls as a cover."

Zhou (also present) laughed. "Especially during elections."

Kissinger laughed.

The next day in the Great Hall of the People, the conversation turned serious. Zhou informed Nixon that China would continue to support

North Vietnam in the war, even if he withdrew American forces completely. Once again, the Chinese premier called for a swifter withdrawal:

> Since the U.S. had decided to withdraw all of its forces from Vietnam and the whole of Indochina, and the U.S. would like to see the region more or less neutral, that is to say, non-aligned, with no particular force occupying that region, then if that is the president's policy and that of your government, I think it would be better to take more bold action. Otherwise, you would only facilitate the Soviets in furthering their influence there. . . . So in this sense the later you withdraw from Indochina, the more you'll be in a passive position, and although your interests is [sic] to bring about an honorable conclusion of the war, the result would be to the contrary. You admitted that General de Gaulle acted wisely when he withdrew from Algeria. In fact, General de Gaulle even withdrew more than two million European inhabitants from Algeria, an action which we didn't dare to envision, and to have withdrawn in such a short space of time. And General de Gaulle encountered great opposition at home.[2]

Nixon lobbied for a decent interval with greater subtlety than Kissinger. America would "let historical processes decide or settle military and political matters in which the issue would be taken to the South Vietnamese," said the president. "We would hope there would be elections."[3] Not that he expected there to be.

As televised spectacle, the summit was a great success. Triangular diplomacy—the hope that Nixon and Kissinger could use China as leverage on the Soviet Union, the Soviet Union as leverage on China, and both as leverage on Hanoi—dazzled the nation. But it didn't stop the North Vietnamese from launching a full-scale invasion of the South a few weeks later.

Sixty-Six Percent for Six Months

On May 8, 1972 (the exact date is important), President Nixon bet on a sure thing. In a live national address from the Oval Office, he announced "decisive military action to end the war." Not only was he sending American bombers to attack military targets in the North, but for the first time, he had ordered the mining of all of its ports.[1]

"It was a difficult decision," he wrote in *No More Vietnams*.[2] In the drawer of the desk in his hideaway office was something that made it a lot easier: the results of a secret poll commissioned by the Republican National Committee three years earlier. Nixon had to keep this one secret.

No president could afford to admit to having a poll done on how to end a war. Two options were clearly unpopular: pressing on until military victory (37 percent in favor); and agreeing to anything to end the war and quickly withdrawing (30 percent). A third got a bare majority: withdrawing over two years while negotiating (51 percent). The fourth option, however, got a massive majority:

> The United States would decide to end the war in Vietnam with a compromise settlement within six months. During those six months, we would try to get the other side to agree to terms that are reasonably favorable to us, such as free elections in South Vietnam under international supervision. To do this we would take necessary military actions, such as [a] blockade of the [North Vietnamese] port of Haiphong and, as a last resort, selective bombing of North Vietnam. No additional U.S. troops would be sent.[3]

Two out of three respondents—66 percent—were in favor of trying to get a compromise settlement out of the North by bombing and blockading it for six months. That's why the date is so important. May 8, 1972, was exactly six months—less one single day—before the election. The only difference between the most popular strategy in the poll and the one Nixon actually announced was the substitution of mines for a blockade of the North's harbors. (Nixon himself called the mining a blockade.)[4] The lesson of the poll was that bombing and mining would be very popular for a half year, so Nixon did both during the half-year period when he most wanted popularity, the one right before Election Day.

Needless to say, when Nixon publicly discussed his reasons for giving the order, he never mentioned its surefire popularity. Battlefield necessity was the public rationale he gave on May 8, 1972: "Five weeks ago, on Easter weekend, the Communist armies of North Vietnam launched a massive invasion of South Vietnam."[5] He wasn't exaggerating. Hanoi had sent every division it had (except for a single reserve) into the South during the Easter Offensive.

The headlines told the story: "Foe Sweeps across DMZ; Saigon Troops Fall Back" (April 2); "Half of Province in South Vietnam Lost to Invaders" (April 3); "N. Viet Invasion Stalls as Allies Launch Huge Counteroffensive" (April 4); "Reds Open Front near Saigon; South Vietnam Fights for Life, Thieu Says" (April 5); "U.S. Sternly Warns Hanoi, Readies New Air Buildup" (April 7); "Laird Confirms U.S. Will Bomb until Reds Withdraw, Negotiate" (April 7).[6] At this point, American officials estimated that the total force North Vietnam was fielding in the South numbered

100,000.⁷ In contrast, South Vietnam's defense forces had grown, with American training and equipment, to more than a million, giving the South a tenfold numerical advantage. But the headlines failed to reassure: "Key Highlands Base Reported Overrun in a Major Offensive by Enemy Tanks" (April 24); "Saigon's Forces Flee in Disorder toward Kontum; Dakto Reported Abandoned—B-52's Pound Enemy to Aid Retreating Troops; Attempt to Regroup; Invaders Have Seized More Than Half of 3 Provinces—4 Die in U.S. Copter" (April 25).⁸

The Easter Offensive had come as no surprise. The North Vietnamese had launched the Tet Offensive in 1968, when LBJ was presumed to be up for reelection, and Nixon expected them to do something like that in his fourth year, too. Intelligence detected a massive troop buildup weeks in advance of the invasion.⁹ The American death toll in the first week of the Easter Offensive was the highest in six months: 10 dead, 33 wounded.¹⁰

Prior to the invasion, Nixon had turned down the military's request to launch air attacks on anti-aircraft missile sites, preferring to wait until the offensive started.¹¹ Once it began, however, he authorized air and sea attacks on the North. On April 26, he announced the withdrawal of another 20,000 American soldiers. All he had to say then about the air and naval attacks was that the ones he had already authorized would continue.¹²

Revealingly, the April 26 speech came nearly three weeks *after* Nixon had Haig draw up a contingency plan to do what the poll said: bomb all military targets in the North and mine its ports. On April 6, when he gave the general this task, Nixon already had a date in mind: "The president ruled that this plan would go into effect May 8 if ARVN"—the Army of the Republic of [South] Vietnam—"showed signs of breaking under the enemy invasion."¹³ The signs were showing well before then, but Nixon held off until his preplanned date.

Between Nixon's April 26 and May 8 speeches, things got much worse: "Fear of Foe Grips People of Pleiku; Hundreds Try to Flee Town in Highlands Expecting the Enemy to Overrun It Soon" (April 29); "Enemy Artillery Batters Quangtri as Ring Tightens" (April 30); "Thousands Flee Kontum in Panic as Enemy Nears" (May 1); "Loss of Quangtri Province Shakes Vietnam's Morale" (May 3); "'It's Everyone for Himself' As Troops Rampage in Hue" (May 4) (the rampaging troops were *South* Vietnamese, runaways who abandoned Quangtri without a fight and menaced their own people in the old imperial city of Hue, according to the *New York Times,* "like armed gangsters—looting, intimidating and firing on those who displeased them"); "Enemy Overruns Base near Pleiku, Killing about 80; South Vietnamese Repulsed in Attempts to Open Roads to Anloc and Kontum; Losses Termed Heavy" (May 6).[14]

With 60,000 American lives at stake, the president told the nation on May 8, "I therefore concluded that Hanoi must be denied the weapons and supplies it needs to continue the aggression." This, he said, was the purpose of bombing and mining the North: to turn the tide of battle in the South by separating the invaders from their supply of arms.[15] Nixon later wrote, "Our best chance of halting the invasion was to take decisive action to stop the shipment of these supplies."[16]

Within the White House, the president gave aides a different rationale. Hitting the North would improve morale in the South, he said, and also improve the American position at the negotiating table. Also, it polled very well. "When they see the enemy invading, the American people support the use of airpower to get Hanoi/Haiphong by 70 percent," Nixon told Haig six days before he announced the bombing and mining. "We polled the goddamn thing. They don't give a shit about negotiations. They don't care." They didn't even care about the nuclear arms limitation treaty he was going to sign in Moscow in a few weeks. Domestic political considerations, the president said, were "the least important, but not to be overlooked." Nixon mentioned one more consideration—getting good press: "Speaking domestically, too, at least when you hit the sons of bitches there, that is the news for two or three days, rather than what the enemy's doing."

"That's right," said Haig.[17]

To the people who answered the pollsters' questions, bombing the North sounded like a great idea. General Abrams, however, objected so strongly that he took a step that ultimately cost him his command of the American armed forces in Vietnam. At the time that Nixon wanted to divert American airpower to the North, thirteen of Hanoi's fourteen army divisions were already in the South, threatening to overthrow the government. Abrams was using the B-52s to stop them. Even if Nixon's plan to bomb the North slowed the flow of supplies, it would take some time before it had an impact on the battlefields of the South. In April and May 1972, Abrams needed American airpower in South Vietnam to avert imminent disaster. As he put it, the president sent the B-52s "away hunting rabbits while the backyard filled with lions." When Nixon ordered a B-52 strike on Hanoi and Haiphong in the first week of May, Abrams did an extraordinary thing. The general overrode the president, saying the bombers were needed in the South to defend its army: "We must stay with them at this critical time and apply the airpower where the immediate effect is greatest."[18] Countermanding an order by the commander in chief was a sign of how seriously Abrams viewed the threat to the South—and the importance he placed on American airpower to ensure its survival.

"Abrams flatly refused to go until I ordered him to do it," Nixon complained to Kissinger.

"Every B-52 strike in the North you've rammed down their throats," said Kissinger. "They didn't want to do it. We've had to order it from here against cables from the Air Force and Abrams telling us not to do it."[19]

The president responded by recalling the general from Vietnam.[20] "This business of having orders countermanded has got to stop," said Nixon. "We've got to get him out of there."[21] The president moved quickly but quietly, kicking Abrams upstairs to Army chief of staff.

All this may come as a surprise to readers of Lewis Sorley's *A Better War: The Unexamined Victories and Final Tragedy of America's Last Years in Vietnam*. In that work, Sorley casts Abrams as an *admirer* of Nixon's bombing of the North. Sorley quotes Abrams as saying: "On this question of the B-52s and the tac air, it's very clear to me that—as far as my view on this is concerned—that this government would now have fallen, and this country would now be gone, and we wouldn't be meeting here today, if it hadn't been for the B-52s and the tac air. There's absolutely no question about it."[22] Sorley took this quote so far out of context as to make the general's position appear to be the opposite of what it actually was. It wasn't until Sorley's next book, *Vietnam Chronicles: The Abrams Tapes 1968–1972*, that he put the quote back in its original context: the general was talking about the importance and effectiveness of American airpower he deployed on the battlefields of *South* Vietnam, not in the North.[23] Sorley has written three books on Abrams, including a full-fledged biography, *Thunderbolt: General Creighton Abrams and the Army of His Times*. Oddly, none of them mentions that Nixon removed Abrams from command, even though this was a pivotal moment in the general's career and life.

The White House cast the invasion as a test of whether Vietnamization had made the South Vietnamese strong enough to stand on their own.[24] Nixon downplayed the American ground role to the point of denial. "There are no United States ground troops involved. None will be involved," said Nixon on April 26; he restated this claim on May 8: "The role of the United States in resisting this invasion has been limited to air and naval strikes on military targets in North and South Vietnam."

Not so. American advisers were embedded with South Vietnamese troops. Abrams saw their ground role as essential to his use of American airpower against the North Vietnamese army. A B-52 needs something to bomb. In the Easter Offensive, that meant that the North Vietnamese had to be gathered in sufficiently large numbers and sufficiently dense concentrations to make a "lucrative target." The way the advisers did that was by giving the North Vietnamese their own "lucrative target"—large numbers of South Vietnamese troops in one place. "Unless the ARVN [South Vietnamese] forces hold on the ground and generate lucrative targets, U.S. and VNAF [South Vietnamese] air power cannot achieve their full effectiveness," Abrams said.[25] The South "Vietnamese, some numbers of them, had to stand and fight. If they didn't do that, ten times the air we've got wouldn't have stopped them."[26] American advisers used their influence to get the South Vietnamese to stand their ground—and they let reporters know that this remained a challenge.[27] Another, equally crucial role of the advisers was coordinating air support—that is, once the North Vietnamese had massed into "lucrative targets," the advisers told the B-52s where to find them. Abrams considered the advisers crucial to the South's survival through the Easter Offensive. "Meeting briefly with Vice President [Spiro] Agnew at Tan Son Nhut Airport on 17 May, he observed that if the South Vietnamese rank and file had fought well when properly led, there were, to his mind, only ten generals in the country who were earning their pay," wrote the military historian William Hammond. "Under those circumstances, the presence of American advisers on the battlefield and the application of unrestricted American air power had been critical in the days following the enemy's offensive."[28]

If it sounds like the South Vietnamese were being used as bait, this approach was originally developed at Khe Sanh in 1968 using American Marines in that role. John Randolph described the tactical breakthrough to readers of the June 30, 1968, *Los Angeles Times*:

> The biggest single tactical problem the allied forces have had in Vietnam is to get the enemy to concentrate in large numbers and then to stay concentrated until allied air and artillery power can hit him.
>
> The idea behind defending Khe Sanh then was that if its Marine garrison looked enough like "bait," the enemy might concentrate large forces near the base in hopes of storming it, thus making themselves a fat target for air and artillery attacks.
>
> This was a sound idea, and in the end worked out almost as hoped for in the beginning—with a punishing defeat for the Communists.[29]

Hanoi took the bait, concentrating at least two divisions outside the garrison. The North lost an estimated 2,600 soldiers during the siege. Most died in air strikes. At the same time, almost 300 American Marines died. "It was a definite allied victory," according to the *Times*.³⁰ In the Easter Offensive, when Abrams applied this tactic, South Vietnamese troops accounted for the great majority of allied casualties, but the North Vietnamese once again took many times more.

In *No More Vietnams,* Nixon presented the South Vietnamese performance in the Easter Offensive as proof positive that Vietnamization worked. He pulled a few fast ones.

> In the spring offensive of 1972, South Vietnam's army had held off the North Vietnamese onslaught without the assistance of any American ground combat troops. [American advisers were ground troops, but technically not combat troops, despite their crucial combat role.] Our senior military commanders in Vietnam and Washington unanimously agreed that the South Vietnamese Army had proved that, if properly equipped and led, it could hold its own against North Vietnam's best troops. [Abrams made it clear that they were "properly . . . led" by American advisers, not so much by South Vietnamese generals.]
>
> We can never know whether the South Vietnamese could have won without the assistance of American air power. [Abrams saw the aerial assault on the North's troops in the South as "the only factor which has prevented a major debacle," and most other observers agreed.] But we know for certain that we could not have won with our air power alone. [Air power required "lucrative targets," and creating those targets required "bait" on the ground.] Vietnamization had worked. Our ally had stopped the spring offensive on the ground [that is, the South Vietnamese stayed in one place long enough for the North Vietnamese to form bomb-able concentrations around them] and our bombing had crushed it.³¹

"American air power inflicted the vast majority of North Vietnamese casualties during the Easter Offensive," Stephen Randolph—a retired Air Force colonel and fighter pilot who at the time of this writing heads the State Department's Office of the Historian—wrote in *Powerful and Brutal Weapons: Nixon, Kissinger, and the Easter Offensive.* "All observers acknowledged the critical role of American advisors, both in solidifying South Vietnamese forces and as conduits to U.S. air and naval power; it was unclear whether the South Vietnamese could perform these functions effectively once the advisors were withdrawn."³²

In other words, the Easter Offensive was no true test of whether the South could survive without American soldiers at their side. The American advisers played two crucial roles in the South's survival at that time, but if Nixon wanted to get a settlement with a ceasefire or just to get the POWs released, he would have to withdraw *all* of the advisers.

The successes of Operation Linebacker (the code name for the May 8 bombing campaign) are famous; its failures, little known. Linebacker produced results. It made real progress toward the goal of choking off the flow of arms and supplies from North to South. It just didn't come anywhere close to achieving it.

The successes were spectacular. The mining (Operation Pocket Money) closed Haiphong completely. Not one ship entered or left the harbor once the mines became active. Linebacker closed rail lines carrying supplies from China. It destroyed the North's small industrial base and knocked out 70 percent of its electrical grid in a couple of months. The statistics were impressive. "For six months," Nixon wrote, "waves of B-52s and F-4s dropped more than 155,000 tons of bombs." Naval bombardment added another 16,000 tons. American air and sea operations cut the flow of Communist supplies in half, and then some.[33] Nixon imposed a high price on Hanoi.

Not high enough, however, to accomplish the mission. "By November," he wrote, "we succeeded in crippling North Vietnam's military effort."[34] Not even close. After four months of bombing and mining the North, the CIA and Defense Intelligence Agency (DIA) assessed the impact it was having on Hanoi's troops in the South. DIA and CIA analysts reached the same conclusion: the North could keep on fighting in the South at its current level even if Nixon continued the bombing and mining *for another two years*.[35] Both intelligence agencies agreed that Linebacker had succeeded in hitting its military targets, the *New York Times* reported, but "it had failed to meaningfully slow the flow of men and equipment to South Vietnam."[36]

The bottom line: for Hanoi to sustain both its economy and the war effort, it needed to import 2,700 metric tons of supplies per day. With its main port closed and Linebacker hitting military targets daily, the North still managed to import 3,000 metric tons per day. In other words, according to the CIA's figures, even with the bombing and mining, Hanoi got 300 more metric tons of supplies per day than were strictly necessary.[37]

The bombing and mining were tactical successes but strategic failures. They weren't "crippling" the North, which continued to get more supplies than it needed to continue the fight.[38]

The strategic failure was predictable—and predicted. Defense Secretary Melvin Laird had warned Kissinger a month in advance that bombing the North would not do the job. Effective interdiction required striking at the source of supplies, Laird wrote on April 6; hitting the distribution system alone wouldn't cut it. The problem was that Hanoi's supplies were produced in the USSR and China—two places Nixon wasn't about to bomb. With the sources of supplies off-limits, Laird wrote, Nixon could only strike at a "diverse and diffused distribution system. The North Vietnamese have demonstrated consistently the ability to substitute new distribution mechanisms for any that are temporarily interdicted."[39]

"If the enemy had one Achilles heel," wrote Nixon in *No More Vietnams*, "it was his supply system."[40] Hanoi's ability to move supplies by truck, boat, cart, and foot made its supply lines more like the multiplying heads of the mythical Hydra: cut one off, and two more would take its place.

Internal government doubts about hitting Hanoi went public during the Easter Offensive when Daniel Ellsberg decided to leak NSSM-1.[41] Most of the coverage focused on the bombing, the issue of the day. At the start of the Nixon administration, the CIA had warned: "The air war did not seriously affect the flow of men and supplies to Communist forces in Laos and South Vietnam. Nor did it significantly erode North Vietnam's military defense capability or Hanoi's determination to persist in the war."[42] The State and Defense Departments agreed with the CIA, but the Joint Chiefs of Staff and General Abrams said bombing the North had been effective during the Johnson years.[43] On April 25, 1972, Sen. Mike Gravel, D-Alaska, tried to enter fifty pages of the study into the *Congressional Record*, saying it proved the president was "pursuing a reckless, futile, immoral policy which he knows will not work."[44]

Since bombing wouldn't work, Laird suggested that the president use his influence with Hanoi's sources in Beijing and Moscow. Nixon made it sound as if triangular diplomacy had closed the aid spigot when he wrote that Hanoi "asked for increased support from its Communist allies. But Moscow and Peking did not man the battle stations."[45] This is misleading. The North, unlike the South, never asked its allies to man the battle stations; the North Vietnamese could do their own fighting (although Chinese troops did fill support roles in the North, freeing North Vietnamese troops for battle). Neither détente with the Soviets nor rapprochement with the Chinese nor the bombing nor the mining dammed the southward flow of supplies from the North. Hanoi's official history of the war claimed that the amount of aid it received actually grew: "The volume of military aid shipped to us by land and sea from fraternal socialist countries"—the Chinese and the Soviets—"and the volume of supplies shipped from the North to South Vietnam in 1972 was almost double that shipped in 1971."[46]

In *No More Vietnams,* Nixon characterized the bombing and mining as the war's "military solution."[47] It was just a political substitute for a military solution.

As much as the May 8 surge failed strategically, it succeeded politically, just as predicted by the secret poll. The pollster Lou Harris reported, "59% of Public Backs Nixon Viet Moves."[48] "The terribly important thing is this," Nixon told Kissinger on May 19, "that it has generally been getting through here due to the polls and everything else that a solid majority on a magnitude of two- to three-to-one of the American people support what we've done. That's point one."[49]

The numbers stayed golden for the six promised months. "Just got Harris's data. It's two-to-one for bombing, two-and-a-half-to-one for bombing," Nixon told Kissinger on September 8. "They want us to be very, very tough, so right now a settlement isn't very much in our interest, unless we can get a ceasefire."[50] Nixon didn't credit his electoral success in 1972 to the summits in Beijing and Moscow. A decade later he was advising President Reagan not to follow in his footsteps by negotiating nuclear arms deals. Polls showed that military action boosts a president's popularity more than diplomacy, Nixon advised Reagan (apparently having overcome his fear that Reagan would blow up the world). After meeting with Reagan in the White House, Nixon dictated a memo: "I pointed out that many people felt my popularity had gone up because of my trip to China. In fact, it had improved only slightly. What really sent it up was the bombing and mining of Haiphong."[51] China and SALT disarmed his critics, but bombs and mines won actual votes.

"One Arm Tied Behind"

Why were bombing and mining so popular? Republicans laid the political groundwork for them throughout the 1960s, with Nixon leading the way. On January 26, 1965, before Johnson sent in American combat troops, Nixon told the Sales Executive Club of New York that the Air Force and Navy could win the war on their own. How? "By naval and air bombing of the Communists' supply routes in South Vietnam and by destroying the Vietcong staging areas in North Vietnam and Laos." Nixon congratulated himself on the courage it took to say the war could be won without American boots on the ground. "The course of action I advocate is one that is not popular in America and would probably not get a vote of confidence in Congress or by a Gallup or Harris poll."[1] (Neither Gallup nor Harris asked Americans if they'd prefer to win the war without having to send in the Army, but who wouldn't?) Johnson ultimately did use air and naval power in North Vietnam and Laos to destroy supply routes and staging areas, but the victory Nixon predicted did not occur. This failure did nothing to diminish Nixon's status as a foreign policy eminence. Hawkish attacks tend to make the attacker look stronger, tougher, firmer than the attacked, even if experience proves them wrong. Sen. Barry Goldwater of Arizona, the GOP's 1964 presidential nominee, soon called for air strikes on Hanoi if bombing supply routes alone didn't "do the trick."[2] Rep. Melvin Laird of Wisconsin, chairman of the House Republican Conference, cast air strikes as a way "to protect American lives and minimize the number of casualties. One such step, already long overdue, is to retarget our bombing raids on more significant targets in North Vietnam."[3] In a major speech in the House on June 16, 1965, Laird called for air strikes on Haiphong harbor. The following month, House Minority Leader Gerald Ford, R-Michigan, said that using airpower and sea power "against significant military targets" was the way to force North Vietnam to settle on acceptable terms.[4] On CBS's *Face the Nation* after Thanksgiving 1965, Nixon called for the policy he would announce seven years later as president: bombing military targets in Hanoi and mining Haiphong. By the end of 1965, Nixon's policy had become the party's, adopted by the Republican Coordinating Committee, whose twenty-eight members included congressional leaders, governors, RNC officials, and all past presidential nominees: "Since it appears that the major portion of North Vietnamese military supplies arrive by sea, our first objective should be to impose a Kennedy-type quarantine on North Vietnam," coupled with "maximum use of American conventional air and sea power against significant military targets."[5]

LBJ's Senate mentor, Georgia Democrat Richard Russell, cracked the formula for a Vietnam victory of the political sort. Continuing the war might ultimately bring success, he told the Senate on March 21, 1966, "but the American people are going to be very unhappy about it, and someone who comes along and says, 'I will go in there and clean this thing up in six months,' will, I am afraid, have some advantage over the senators who say, 'Let's play this thing along for ten or twelve years.'" That day Russell came out for mining Haiphong.[6]

As air and naval bombardment of the North failed to yield victory in the 1960s, a stabbed-in-the-back myth began to emerge, blaming the failure on restraints and restrictions that allegedly kept America from using its power effectively. Gen. Curtis LeMay, the former Air Force chief of staff, told *Human Events* in 1967: "We always set up a bunch of arbitrary rules so that we can't use the power we are capable of bringing to bear. Consequently we have a long, drawn-out war and I think the loss of life is much greater in the long run than it would be were we to use the force necessary to stop it early in the game." LeMay noted that the United States had already used more explosives on North Vietnam than Germany had dropped on England in the Battle of Britain, "but we haven't gotten much in the way of results." Bombing slowed down the North's movement of supplies, LeMay said, but didn't stop it. This didn't shake LeMay's faith in bombing; he just thought it needed to be used against other, better targets: "We have to go back a little bit farther to the source of supply. I would close the harbor of Haiphong and certainly prevent anything from getting into the country." The United States could bomb the harbor or blockade it, said LeMay. The possibility that closing Haiphong would *not* prevent supplies from getting into the North—that the Communists would change their supply routes as they had before— just didn't come up then.[7]

Gov. Ronald Reagan, too, blamed the lack of victory on restrictions. "There [is] still a list of targets that are not open to bombing by our forces," he complained during a September 12, 1967, press conference. Reagan saw the failure of the escalating air and sea bombardment of the North merely as an indication that LBJ had not escalated fast enough: "The war might have ended, because doing it all at once might have brought the enemy to the bargaining table," Reagan said. He favored further escalation "to win the war as quickly as possible." The possibility that Hanoi might be able to handle faster or greater escalation just didn't come up then, either.[8]

Republicans and conservatives had a well-crafted narrative of the Democrats' failure in Vietnam before Nixon entered the White House. It blamed the Democrats for using too little force, or using it too late, or imposing restraints and restrictions that prevented it from being used against the right targets. Reagan came up with the perfect metaphor when, as president, he told Vietnam veterans, "You fought as bravely and as well as any American in our history, and literally with one arm tied behind you. Sometimes two."[9] The metaphor, so vivid and clear, spared Reagan the burden of identifying which targets needed to be hit, which restrictions and restraints needed to be lifted, to reap victory in Vietnam. It also avoided identifying who did the alleged tying, why, or how, sparing Reagan the additional burden of providing evidence to back up his accusation. As a case, it was weak; as an expression of the stabbed-in-the-back myth, which depends on the suppression (or manufacture) of evidence, it was perfect. Reagan's words also contained a kernel of truth: defeat in Vietnam truly was not the fault of the American soldiers sent to Vietnam. It was the fault of the politicians who sent them there without ever coming up with a strategy to win the war.

"Why Does the Air Force Constantly Undercut Us?"

Nixon exploded when he heard that even after May 8, 1972, Adm. John McCain Jr. had complained about bombing restrictions.

"Apparently, the attitude that I understood was that there's some remorse over the fact that there have again been limits put on bombing," Vice President Spiro Agnew reported to Nixon on May 19 after returning from a trip to Asia.

"Now what in the hell is—"

"Those sons of bitches of the Air Force!" said Kissinger, cutting off the president. "If we hit this once more, I'm going to recommend to the president to fire that bastard of a chief of staff of the Air Force. There are no limits on the bombing."

"Who'd you get this from?" Nixon asked. "Abrams?"

"McCain," said Agnew.

As commander of American armed forces in the Pacific arena, Admiral McCain was Abrams's superior. Nixon had considered making McCain his "supreme allied commander" in Vietnam to take control of the bombing and mining away from Abrams. "I have more confidence in the admiral than I have in Abrams," Nixon told Haig on May 5. "Actually, you know, he's got a son who's a POW."

"That's right," said Haig.

"And he is a little hard-line son of a bitch. Now, I know he'll make the mining work. But for example, with regard to the allocation of bombers and everything else, I want him to do it rather than Abrams."

Haig wasn't sold on McCain: "I don't have much confidence in his brainpower. His heart is great."[1]

After hearing from Agnew, Nixon summoned the chairman of the Joint Chiefs of Staff, Adm. Thomas Moorer, for an Oval Office dressing-down. Waiting, Nixon fumed: "The Air Force just dropped the ball in a miserable way. They aren't worth a goddamn. They aren't worth a goddamn. Incidentally, I want the head of that son of a bitch at the Air Force [Gen. John] Ryan today. He's out. Out, out, out," said the president. "They've got total freedom to bomb anyplace they want."

"Of course," said Kissinger.

Moorer entered.

Nixon began an angry monologue: "Ryan has got to get off his goddamn ass or he's out. I'm tired of him anyway. He's a soft man. I mean, of course, he should be, I mean, his son killed and all that sort of thing. But let me say, you know and I know, I have ordered that goddamn Air Force time and time and again to do anything, and they can't bomb, because they say these 4[,000]—they need a 4,000-foot ceiling." The cloud ceiling affects flight visibility. "You know and I know that they do not have restrictions. The only restriction they've got is the one within ten miles of Hanoi *at the present time*." (American bombers were ordered to steer clear of Hanoi starting May 20 so they wouldn't accidentally bomb the Soviet embassy while Nixon and Kissinger were attending the Moscow summit.) Nixon railed:

President Nixon: The Air Force didn't do a *goddamn thing* for the last three days, as you know. Not one goddamn thing in North Vietnam, because the little bastards were afraid that they might not—they might lose a plane because they couldn't see. I am tired of this bullshit! It's been in every paper in this town. They're telling the vice president this. As you know, they're whining around. Now, never have they had the backing they've got today, and I want the military to shape up or there's going to be a new chief of staff all up and down the line.

The president slammed his desk as he said, "Now you go take care of it right now. Is that clear?"

"Yes, sir."

"Now, get off your ass. Now, I want you to get that son of a bitch Ryan on the phone. I want you to get McCain on the phone. Are you restricted?"

"No, sir," said Moorer, noting the exception for the Moscow summit.

"Yeah, yeah. Are you bitching about it?"

"No, sir."

"What the hell is the matter with these people? Why are they whining? Because they're afraid to go in and do the job that they've been ordered to do? What in the hell is the matter with them now?"

Agnew intervened: "I must say that, for example, there was a spot where they had taken out some very important rail that if they had been able to go back and really take it out, they were told they couldn't go back, so . . ."

"Now that is absolutely false," said Nixon. "That is absolutely false. Not only have they been able, but I have watched every day, and the goddamn air force *doesn't* go back, because they're afraid that the weather isn't good enough. They've got to have 5,000-foot ceilings. The goddamned Israelis fly at 1,000-foot ceilings. Now, tell them to get off their goddamn ass and do the job. And I—like, for example, I want some [B-]52s to hit them. Oh, no, Abrams needs them in the South. All right, fine, we'll keep them in the South, but for Christ's sakes, why does the air force constantly undercut us and bitch when they've never been backed as they're backed today? Tell them to do the job. Now, incidentally, I really mean it. Ryan is going to have a resignation on this desk. I'll fire his ass out of here unless he gets some discipline in that damn outfit. Is that clear?"

"Yes, Mr. President," said the JCS chairman.[2]

Nixon and his fellow Republicans, along with conservative Democrats, had turned "restrictions" and "restraints" into political curse words. He would not allow the military to apply these terms to *his* bombing.

The Appearance of Success

As Nixon's bombing in the North improved his numbers in the polls, General Abrams's bombing in the South improved conditions on the ground. The president increased the airpower available for all purposes in Southeast Asia, so Abrams had all the B-52s he needed to pound North Vietnamese divisions even after Operation Linebacker began. Abrams found a few kind words to say about Linebacker, calling the results it produced "very substantial." Nevertheless, he cautioned that "it is not possible to lose the war in the North, but it is still possible to lose the war in the South, and we must not turn loose of this until the job is done."[1] The headlines began to get better in May 1972: "S. Vietnam Troops Turn Back Red Attack at Kontum" (May 15); "S. Viet Units Break Long Kontum Siege, Push on Quang Tri" (July 1); "Saigon Reports Its Troops Enter Quangtri Capital" (July 5); "Saigon Reports Repulsing an Enemy Attack on a Pleiku Outpost" (September 6).[2]

Nixon now had a coincidence he could take advantage of: since Operation Linebacker had gotten under way right around the same time that Abrams began grinding the Easter Offensive to a halt, Nixon could falsely claim that bombing the North had *caused* success in the South.[3] *Post hoc, ergo propter hoc* is a fallacy so ancient that the first people to skewer it wore togas and spoke Latin, but that doesn't stop it from continuing to work in modern politics.[4] Americans had been hearing politicians say for years that restrictions on bombing the North were preventing success in Vietnam. When Nixon lifted those restrictions, and Abrams achieved some limited success in stopping the Easter Offensive, these developments fit into the conservative narrative. It *looked* as if the bombing and mining were working, even as they were failing.

Despite the headlines, the North Vietnamese would end the Easter Offensive in control of more territory in the South than they held at the start. With a ceasefire-in-place, this would give them a platform in the South that they could use to launch another bid to overthrow the Saigon government after Nixon withdrew American forces.

"Any Means Necessary"

Looking back on the bombing and mining a decade after his presidency ended, Nixon wrote: "My only regret was that I had not done it earlier."[1] In fact, he had made plans to do it three years earlier. During his first year in office, Nixon went so far as to secretly give Hanoi an ultimatum. To deliver it, Nixon recruited a leader of the French Underground who had led France's unsuccessful negotiations with Ho Chi Minh at the end of World War II, Jean Sainteny.[2] Through Sainteny, Nixon set a deadline—November 1, 1969—for a "positive outcome" to the Paris peace talks.

> But if, however, by this date—the anniversary of the bombing halt [that LBJ ordered in 1968 to get the peace talks started]—no valid solution has been reached, he will regretfully find himself obliged to have recourse to measures of great consequence and force.... Regardless of public opinion or opposition, Mr. Nixon is determined to bring this war to an early conclusion. He totally rejects continued talking and fighting. If this diplomatic approach fails, he will resort to any means necessary.[3]

The message was simple: settle or else. Nixon held close to his vest the precise military cards he planned to play while hatching sundry schemes to make the Communists fear the worst. He made comments to Republican senators that he knew would leak; within days, the columnists Rowland Evans and Robert Novak reported that Nixon might not only hit Hanoi's docks but order an actual invasion of the North.[4] He staged a phone call while Kissinger was meeting with Soviet ambassador Anatoly Dobrynin to say, in the hoary Washington cliché, that the train had left the station. The hope was that Hanoi's chief supplier of military and economic aid would urge the North to get on board with a settlement.

Nixon's most elaborate stratagem was intended to give Moscow the impression that Nixon just might go nuclear. In this, Nixon thought himself to be merely following in Eisenhower's footsteps. Eisenhower claimed that during the negotiations to end the Korean War, "in India and in the Formosa Straits area, and at the truce negotiations at Panmunjon, we dropped the word, discreetly ... that, in the absence of satisfactory progress, we intended to move decisively without inhibition in our use of weapons." In explaining how the Korean armistice came about, Nixon said that "it was the Bomb that did it." (Historian Jeffrey Kimball disagrees: "Nothing in the diplomatic front or on the battlefield, where the Chinese continue to launch attacks and counterattacks up to the signing of the armistice in July 1953, indicated that they were intimidated by nuclear threats or by the U.S. Air Force's massive conventional bombing of the North Korean dam-irrigation system in May.")[5] As his November 1 deadline approached, Nixon ordered a worldwide nuclear alert. The Strategic Air Command (SAC), whose bombers and intercontinental missiles were the most conspicuous elements of America's nuclear arsenal, placed B-52s, Phantoms, and other aircraft on a heightened state of readiness. At Adm. John McCain's suggestion, the Navy engaged in surveillance of Soviet ships en route to Hanoi and stepped up patrols by submarines armed with Polaris nuclear missiles. By October 27, SAC had six bombers armed with nuclear weapons flying continuously over the frozen Arctic wasteland, where they were certain to trigger Soviet early-warning systems.[6]

Nixon's November 1, 1969, deadline came and went. The North Vietnamese had not provided the "positive outcome" he had demanded in Paris. Instead, they called his bluff.

What "measures of great consequence and force" did Nixon take in response? None. He decided against carrying out his threat. In his memoirs, he put the blame squarely on antiwar demonstrators. They had organized two massive, nationwide days of protest—one on October 15, the other on November 15—that together "destroyed the credibility of my ultimatum to Hanoi."[7] This assumes it had credibility in the first place. Nixon had already demonstrated his unwillingness to carry out the ultimatum that his predecessor, Lyndon Johnson, had secretly made against Hanoi before halting American bombing of the North on October 31, 1968. Johnson had his negotiators repeatedly warn Hanoi's that he would resume the bombing if it (1) violated the DMZ, (2) failed to engage in productive talks with South Vietnam, or (3) shelled civilians in South Vietnamese cities. Hanoi had inarguably violated some of those conditions and arguably violated all of them months earlier. Yet Nixon had not made good on Johnson's threat. That told Hanoi more about the credibility of Nixon's ultimatum than any antiwar demonstration.

Nixon also wrote in his memoirs that he worried that "American public opinion would be seriously divided by any military escalation of the war."[8] He omitted mention of the poll in his drawer indicating that opinion would divide sharply *for* escalation: 66 percent for bombing and mining to only 26 percent against.[9] At the time, however, Nixon told Kissinger that he "would just as soon have them demonstrate against the plan. If we went ahead and moved, the country is going to take a dimmer view after the move than before." Even Defense Secretary Laird, who apparently had not seen the poll either, estimated that public support for the surge would remain "relatively high" for three months. Nixon had excellent reason to think he could maintain support for bombing and mining the North for six months. Yet he didn't chance it—despite recognizing the risks involved in revealing himself to the Communists as a maker of empty threats: "I knew that unless I had some indisputably good reason for not carrying out my threat of using increased force when the ultimatum expired on November 1, the Communists would become contemptuous of us and even more difficult to deal with."[10]

Ultimately, Nixon wrote, three things made up his mind: (1) American casualty rates "had been reaching new lows"; (2) Ho Chi Minh's death "might have created new opportunities for reaching a settlement"; and (3) the British guerrilla warfare expert Robert Thompson said that in two years the South Vietnamese would be ready to defend themselves. However, (1) American casualties had been declining steadily each month since May 1969; this wasn't a new development in October; (2) the death of Ho Chi Minh was not about to *decrease* North Vietnamese nationalism; and (3) Thompson's faith in the potential of Vietnamization had to be weighed against the consensus of General Abrams, the JCS, the CIA, the State Department, and the Pentagon that the training program would never make Saigon capable of surviving without American troops.[11] Nixon's own disbelief in Vietnamization, manifest on his tapes and in his refusal to risk withdrawing the last American troops until his reelection was secure, undermines the explanation he gave readers for his decision to back off of his ultimatum: "I began to think more in terms of stepping up Vietnamization while continuing the fighting at its present level rather than of trying to increase it. In many respects Vietnamization would be far more damaging to the Communists than an escalation that, as Thompson had pointed out, would not solve the basic problem of South Vietnamese preparedness."[12]

Nixon had a better reason for backing off. The risk was high that bombing and mining would fail. Kissinger nailed the problem on October 2, 1969, in a sixty-one-page top secret memo to the president: "We must accept from the outset that Hanoi will be an extremely tough nut to crack, and that the North Vietnamese leaders may well conclude that having held out this far, they can do so sufficiently longer to leave us no choice but to back off." They had already withstood four years of American bombing under LBJ. The North had enough supplies prepositioned in the South to sustain its troops for months. Nixon could bomb more targets, and add mining to the mix, but Hanoi might just adapt to new circumstances as it had before (and would do again after May 8, 1972).

As Kissinger saw it, "our main problem will not be the specifics of what Hanoi does against us but whether or not the North Vietnamese leaders will move toward a compromise within an acceptable time frame." What if Nixon bombed and mined the North for six months and Hanoi still didn't settle? The secret poll showed high support for the surge as a way to end the war, not to continue it a few more years. Yet Kissinger had warned that if he started bombing and mining, he had to stick with it: "To attempt this course and to fail would be a catastrophe. It must therefore be based on a firm resolve to do whatever is necessary to achieve success. Since we cannot confidently predict the exact point at which Hanoi would be likely to respond positively, we must be prepared to play out whatever string necessary."[13] At the end of six months of bombing and mining, Nixon would still have had two and a half years to go before Election Day—more than enough time for voters to realize that bombing and mining weren't working.

Those were some of the problems that would arise if the surge didn't work; others would arise if it did. Thompson hinted at them when he mentioned the South's lack of preparedness. Even if Hanoi had agreed to a settlement with Nixon in 1969 or 1970, that wouldn't necessarily mean abandoning its goal of taking over the South. Suppose Nixon had somehow persuaded Hanoi to agree to mutual withdrawal. It's a truism that you won't win at the negotiating table what you can't win on the battlefield, but suppose that Nixon, somehow, managed to do just that. Once he withdrew American troops halfway around the world, Hanoi could start infiltrating soldiers back into the South once again. Knowing that South Vietnam's army could not withstand the North's on its own, Nixon would face some bad choices. If he didn't intervene with American ground troops, Saigon would fall, and he would get the blame. If he did intervene with American ground troops, America would be back at war and as far from "peace with honor" as the day he took office. (The problems that would have arisen if Nixon had intervened with American air and naval power, but not ground troops, are the subject of later chapters of this book; suffice it to say for now that air and naval power had never proven to be enough to stop North Vietnam in the past.)

Backing off from the ultimatum enabled Nixon to avoid potential political problems, but it created a geopolitical one: the Communists realized he was a paper tiger. Two days after his deadline passed, Nixon delivered a nationally televised address on Vietnam announcing... no new military action. Ambassador Dobrynin soon expressed Moscow's scorn to Ambassador Llewellyn Thompson, saying that "he did not understand why there had been such a big buildup beforehand."[14]

Politically, however, Nixon's November 3, 1969, speech was the most successful of his first term. It was the Silent Majority speech, and it succeeded as much by what it didn't say as what it did: "I believe that one of the reasons for the deep division about Vietnam is that many Americans have lost confidence in what their government has told them about our policy. The American people cannot and should not be asked to support a policy which involves the overriding issues of war and peace unless they know the truth about that policy."

Of course, he did not tell the American people the truth about his policy. He kept secret the fact that he had made an ultimatum, that the Communists had called him on it, and that he had backed down (although none of this was a secret from Moscow or Hanoi). Shrewdly, Nixon released the text of his letter to Ho—but not of the threat that Sainteny delivered with it. Just as shrewdly, he released Ho's reply, which ignored the threat completely. Ho's reply was evidence of the threat's failure, but Nixon presented it as evidence of Hanoi's intransigence.

The Silent Majority speech laid the foundation for the stabbed-in-the-back myth Nixon would create to blame the fall of Saigon on his critics rather than himself. Although Vietnamization was a slow path to defeat, Nixon cast his strategy as the only way to *avoid* defeat, claiming that "withdrawal will be made from strength and not from weakness. As South Vietnamese forces become stronger, the rate of American withdrawal can become greater."

> My fellow Americans, I am sure you can recognize from what I have said that we really only have two choices open to us if we want to end this war.
>
> I can order an immediate, precipitate withdrawal of all Americans from Vietnam without regard to the effects of that action.
>
> Or we can persist in our search for a just peace through a negotiated settlement if possible, or through continued implementation of our plan for Vietnamization if necessary—a plan in which we will withdraw all of our forces from Vietnam on a schedule in accordance with our program, as the South Vietnamese become strong enough to defend their own freedom. I have chosen this second course.

Nixon rhetorically divided the nation into two groups—those who were for him and those who were for losing:

> I recognize that some of my fellow citizens disagree with the plan for peace I have chosen. Honest and patriotic Americans have reached different conclusions as to how peace should be achieved.
>
> In San Francisco a few weeks ago, I saw demonstrators carrying signs reading: "Lose in Vietnam, bring the boys home."
>
> Well, one of the strengths of our free society is that any American has a right to reach that conclusion and to advocate that point of view. But as President of the United States, I would be untrue to my oath of office if I allowed the policy of this nation to be dictated by the minority who hold that point of view and who try to impose it on the nation by mounting demonstrations in the street.

If you didn't support Nixon's strategy—if you thought, for example, that there was no way Vietnamization would make the South Vietnamese army strong enough to take the place of America's in the war—then, according to Nixon, you would be to blame for defeat.

> Let us be united for peace. Let us also be united against defeat. Because let us understand: North Vietnam cannot defeat or humiliate the United States. Only Americans can do that.[15]

Nixon had demonstrated his ability to lose the war and humiliate the nation on his own, without the assistance of his critics, by adopting a strategy of withdrawing all American troops from Vietnam and backing down from his ultimatum. But NSSM-1 and the ultimatum remained secret, allowing Nixon ample time and opportunity to set up his critics to take the blame for the consequences of his actions.

"A Russian Game, a Chinese Game and an Election Game"

Nixon milked May 8 for every drop of drama. He played up the possibility that the Soviets might call off the summit in Moscow. "Most of the members of Congress, my cabinet, and my staff shared the view that the summit would probably be off," wrote Nixon. "Kissinger and I agreed that there was a good chance the Soviets would cancel or postpone our summit meeting."[1]

Actually, the president and national security adviser calculated correctly quite early in the Easter Offensive that the Soviets would keep the summit. On April 3, the day after offensive began, Kissinger said, "In a way, it's a godsend. We should give them a tremendous punishment." Kissinger proposed calling in Ambassador Dobrynin and threatening that Nixon, in retaliation, would refuse to sign the Berlin treaty.

"That's right," said Nixon. "Having in mind the fact that [unclear] we still want to drive a hard bargain on the summit. Oh, they want that summit."

"Mr. President, they can no more afford to not have that summit—"

"They can't trade Vietnam for this," said Nixon. This initial judgment proved correct. If Nixon could hold a summit while Soviet-made weapons were killing Americans, Brezhnev certainly could.

"I think we could play this into an end of the war," Kissinger said of the offensive.

"I think you're right," said Nixon.

"If the ARVN collapses, we've done everything we can, Mr. President," said Kissinger.

"If the ARVN collapses, a lot of other things collapse around here. If we were going to collapse, we had to do it a year ago. We can't do it this year, Henry."

"God, they're not going to collapse," said Kissinger. "That's why we've got to blast the living bejeezus out of North Vietnam."

"Let's don't talk about if the ARVN collapses, you know what I mean, [unclear]. That's fine with regard to this, but we're playing a much bigger game," said Nixon. "We're playing a Russian game, a Chinese game and an election game. And we're not going to have the ARVN collapse."[2]

The way Nixon told it later, the bombing and mining made the summit possible. His account in *No More Vietnams* made it seem as if the bombing in the North had had an immediate effect on the war in the South:

> If we had not acted, we might have had to go to Moscow while Soviet-made tanks were rumbling through the streets of Hue and Saigon. We would have been in an intolerable position of weakness. Brezhnev would have assumed that if I could be pushed around in Vietnam, I could also be pushed around in Moscow.
>
> Our diplomacy with Moscow and Peking had turned the tables on Hanoi. It had been an article of faith within the Kennedy and Johnson administrations that making a decisive military move against North Vietnam risked the intervention of China and the Soviet Union. That now changed: Hanoi was fearful that its allies might use their leverage to intervene on the side of its enemy.[3]

This was a freestyle rewriting of history. May 8 was no "decisive military move against North Vietnam." Like his Democratic predecessors, Nixon avoided destroying the North Vietnamese government, the one thing he and they agreed would bring on Chinese intervention. The bombing and mining, far from demonstrating Nixon's strength, provided him with political cover to make a crucial, secret concession.

Even while B-52s were dashing the Communists' hopes for immediate military victory on the battlefields of South Vietnam, Nixon and Kissinger were raising the Communists hopes for victory after a "decent interval" behind the scenes at the Moscow Summit. "Dr. Kissinger told me that if there was a peaceful settlement in Vietnam you would be agreeable to the Vietnamese doing whatever they want, having whatever they want after a period of time, say 18 months," General Secretary Leonid Brezhnev said on May 24. Eighteen months—a year or two—was the same figure Kissinger had used with Zhou the first night they met. "If that is indeed true, and if the Vietnamese knew this, and it was true, they would be sympathetic on that basis," Brezhnev said. "Even from the point of view of the election in the United States I submit that the end of the war at this particular time would play a positive role, whereas escalation will not." The general secretary of the Soviet Communist Party did not study public opinion as closely as the American president. "As for sending in new waves of bombers against Vietnam, they cannot solve the problem and never can." Brezhnev was right about the bombing's strategic impact, wrong about its political impact.

Nixon didn't say anything one way or the other about the eighteen-month period.[4] Kissinger got more specific with Soviet foreign minister Andrei Gromyko on May 27. Kissinger still euphemistically divided the issues into "military" (American withdrawal, release of POWs, a ceasefire-in-place for the Vietnamese) and "political" (how the Vietnamese would determine who governed South Vietnam), but he emphasized Nixon's intention to stay out of Vietnam after a settlement of the "military" issues. "We would then guarantee, except for economic and military aid, to keep our hands out of it; we would be neutral in the political process," said Kissinger. "We would not trick North Vietnam. If we would, then the fighting would start again. That is not in our interest. All we ask is a degree of time so as to leave Vietnam for Americans in a better perspective."

"My impression sometimes from the president and Dr. Kissinger, the official position of the United States is that it is impossible to leave Vietnam to some kind of Communist or Socialist government," said Gromyko.

"What we mean is that we will not leave in such a way that a Communist victory is guaranteed. However, we are prepared to leave so that a Communist victory is not excluded, though not guaranteed," said

Kissinger. "Our position is we want a political solution which does not guarantee a Communist victory, but also, we emphasize, that does not exclude it."

"That is official?" Gromyko asked.

"You can communicate this to the North Vietnamese," said Kissinger.[5]

"It Could Be a Bit Longer"

In his best-selling, multivolume memoirs, Kissinger let historians of the Vietnam War know that they should carefully examine the transcripts of his June 1972 trip to Beijing. He did so by writing, "My visit brought no new developments on Vietnam."[1]

By the time Kissinger arrived in Beijing, the Chinese premier was finally ready to discuss a "decent interval." For three days Zhou Enlai probed the depths of the Occidental need to save face.

If, after American troops and POWs went home, "civil war again breaks out in Vietnam, what will you do?" the Chinese premier asked on June 20 in the Great Hall of the People. "It will probably be difficult for you to answer that."

On the day they had met almost a year earlier, Kissinger had answered plainly enough: America would not intervene if there were a ceasefire for eighteen months—that is, a so-called decent interval.

Now the national security adviser was more circumspect: "It is difficult for me to answer partly because I don't want to give encouragement for this to happen." If a month after American withdrawal "the war starts again, it is quite possible we would say this was just a trick to get us out and we cannot accept this. If the North Vietnamese, on the other hand, engage in a serious negotiation with the South Vietnamese, and if after a longer period it starts again after we were all disengaged, my personal judgment is that it is much less likely that we will go back again, *much* less likely."

"You said this last year, too," said Zhou.[2] (Not quite. At their first meeting, Kissinger spoke in absolutes, not probabilities: "if it is overthrown after we withdraw, we will not intervene.")[3]

Kissinger didn't disagree: "Last year if they had accepted our proposal, it would now have been a year. If the North Vietnamese could transform this . . ."

"You said last year after you have withdrawn and the prisoners of war have been returned then as to what happens then, that is their affair. In principle you mentioned that."

"In principle we are attempting to turn . . . it, of course, depends on the extent to which outside countries intervene. If one can transform this from an international conflict in which major world powers are involved, to a local conflict, then I think what the Prime Minister said is very possible. But this is our intention and since we will be making that policy, it is some guarantee."[4]

The next day, Zhou pressed a bit harder on the question of postwithdrawal intervention.

"If the North Vietnamese accept a ceasefire, we will withdraw all our forces in return for our prisoners," said Kissinger.

"Does that mean your air and naval forces will also withdraw?" asked Zhou.

"Yes."

"So then since you will have by that time withdrawn all your forces and have all your prisoners of war repatriated, then if the political issue cannot be solved and a civil war breaks out again, you shouldn't go back to take care of that," said Zhou.

Kissinger used two synonyms for "decent interval" in his reply: "Mr. Prime Minister, you have more experience in international affairs than I so you know there are certain situations in which it is very difficult to give a formal answer, because one does not want to create a legal obligation for what may be taken care of by reality. I believe if a sufficient interval is placed between our withdrawal and what happens afterward, that the issue can almost certainly be confined to an Indochina affair; and if there is no other outside intervention." (North and South Vietnam were both within French Indochina, of course, and the United States, China, and the Soviet Union were not.) "From your own analysis of the American situation," Kissinger said, "it should be self-evident that in a second term we would not be looking for excuses to re-enter Indochina. But still it is important that there is a reasonable interval between the agreement on the ceasefire, and a reasonable opportunity for a political negotiation." Kissinger was cagey about what would happen after the "sufficient" or "reasonable interval."

"So the outcome of your logic is that the war will continue?" asked Zhou.

Kissinger used another synonym for "decent interval" in his reply: "No, the outcome of my logic is that for the time period . . . I am not trying to win debating points, because I agree with the Prime Minister that we have a difficult problem to settle. The outcome of my logic is that we are putting a time interval between the military outcome and the political outcome."

This was Kissinger's old euphemism: the "political" issue was who would rule South Vietnam—an issue that all parties knew would be resolved militarily. But this misleading distinction allowed Nixon and Kissinger to say they would resolve the military issues of complete American withdrawal, release of American POWs, and a ceasefire-in-place for the Vietnamese, while leaving the "political" issue of determining the South's future up to the Vietnamese. "No one can imagine that history will cease on the Indochina peninsula with a ceasefire. And I believe that if the North Vietnamese had confidence in themselves they should have a better chance this way than through a continuation of the war," said Kissinger. "The Prime Minister also said that the issue is that we do not want to give up a certain government. That is not correct. What we will not do is ourselves to overthrow the government. We will agree to an historical process or a political process in which the real forces in Vietnam will assert themselves, whatever these forces are. Why should we be afraid of socialism in Vietnam when we can live with communism in China?"

Zhou laughed and said, "That is the point that I don't understand." The coalition government proposed by the North would bring Communism to the South, too, but with far less bloodshed.

"The reason we cannot accept this government of national conciliation is because its objective consequence will be to overthrow the existing government and bring into power—it is a very thinly disguised formula for bringing into power the DRV [Democratic Republic of Vietnam, i.e., North Vietnam]. And therefore we believe that the most rapid way of ending the war would be to concentrate on the military issues and permit us to disengage from Indochina, and after that permit the local forces to work it out, either through negotiations or other means." Kissinger didn't need to specify what those "other means" were.

"You mean the local forces in Vietnam?" asked Zhou.

"Yes."

"But if the issue cannot be solved at the present existing negotiation table, then the local forces will continue in conflict," said Zhou. "This can almost be said to be certain."

"Then," Kissinger said, "at least the outside forces will be disengaged."[5] This was what Zhou needed to know. With America out of the war, the North could handle the South.

On the last day of Kissinger's trip, close to midnight, the Chinese premier gave him the answer he'd been seeking for more than a year. Zhou could not speak for North Vietnam, but he could for China: "We cannot impose our thinking on a small country. If one were only proceeding from our thinking, our thinking would be like yours," Zhou said. "You go your way, and as for Vietnam affairs, you cannot meddle in them anymore. As you said, one month would be too short for the fighting to break out again. It could be a bit longer."[6] Beijing, like Moscow, was willing to delay Communist victory in Vietnam for a "decent interval."

The following month, Zhou welcomed another guest, this one the third most powerful official in Hanoi's Politburo. Le Duc Tho handled the critical negotiations with Kissinger. He was Zhou's fellow revolutionary, having spent more than ten years in a French colonial prison after founding Indochina's Communist Party. He knew how to deal with setbacks. "We do not recognize Nguyen Van Thieu, as he is a puppet of the U.S.," said Zhou on July 12. "Yet we can recognize him as a representative of one of the three forces in the coalition government." Hanoi was still demanding a coalition government of Communists, anti-Communists, and neutrals. Nixon wouldn't accept a coalition, but he and Kissinger were willing to have those three elements in an election commission for South Vietnam, provided it was sufficiently toothless. "In case negotiations among the three forces fail, we will fight again," said Zhou. He used "we" fraternally, referring to the Communist side.

"We are asking Thieu to resign," said Le Duc Tho. "If he does not, we will not talk with the Saigon government."

"If he does, who will replace him?" asked Zhou.

"We are ready to talk with anyone."

"That also means Thieu's policy without him," said Zhou. "Chairman Mao has also spent much time talking with me on the question of a tripartite government. He told me to talk with you on this issue. We also have experience on this issue. A coalition government could be established, but we later had to resume fighting. The question is to play for time with a view to letting North Vietnam recover, thus getting stronger while the enemy is getting weaker."[7]

The Democrats

The Democratic general election campaign of 1972 is best remembered as a series of cautionary tales with clear, straightforward morals—a checkoff list of political don'ts. Don't allow convention delegates to nominate just anyone for vice president (the nominees included Mao Zedong, the CBS substitute anchorman Roger Mudd, and the television character Archie Bunker). Don't let the resulting delays push back the acceptance speech of the presidential nominee to 2:48 a.m., prime time in only one state, Hawaii.[1] Above all, don't nominate anyone for vice president without exhuming every potential embarrassment from that person's past and assessing its potential as a front-page story. If the front pages reveal that the vice-presidential nominee underwent electroconvulsive therapy (which he failed to disclose) as treatment for clinical depression (which he also failed to disclose), don't permit the presidential nominee to declare "1,000 percent" support for the embattled running mate shortly before seeking a replacement.[2] Don't deplete journalistic reserves of synonyms for disaster. Spring was the season of Republican sabotage; in summer, the Democrats sabotaged themselves.

It might seem unnecessary, even unkind, to draw attention to a previously unnoticed flaw in a campaign noted largely for its flaws. But McGovern was already behind Nixon before his summer of sadness. He was losing the general election even while he was winning primaries. The signature issue of his campaign was the central issue of *the* campaign: Vietnam. His landslide loss was not the climactic pratfall at the end of a row of banana peels. It was the direct result of two strategies, his and Nixon's, based on two competing visions of the politics of the war—on two competing stories about Nixon and Vietnam. Tragically for American voters, neither of their stories was true. McGovern misled by accident, Nixon by design, but only one of their stories appeared to fit the facts. (Neither story actually did fit the facts, but Nixon's at least appeared to do so.)

It would have been different if the 1972 Democratic convention had nominated someone who told Ted Kennedy's story. But of all the leading Democrats, only Kennedy accused Nixon of using Vietnamization as a cynical excuse to keep the war going so he could end it closer to Election Day. Only Kennedy charged Nixon with putting politics above the lives of American soldiers. Only Kennedy rejected the prevailing liberal view of Nixon as the unyielding defender of South Vietnam's independence, seeing him instead as a political opportunist seeking merely to delay defeat. The story Kennedy told fit the facts that would emerge in the final days of the campaign better than the stories told by McGovern or Nixon. Failing to send a nominee into the fall campaign armed with the Kennedy story was the biggest strategic failure of the 1972 Democratic convention. It forfeited an opportunity to seize control of the narrative.

If other Democrats had picked up on the story Kennedy was telling about Nixon and added their own voices to it, voters might have seen Nixon's slow, four-year, gradual withdrawal of American troops from Vietnam for what it was: a way to make it look, right before Election Day, as if Vietnamization had made South Vietnam capable of its own defense. Instead of being impressed or relieved, they might have been outraged that a president had put his reelection above the lives of their children. The story that Kennedy told about Nixon was compelling. It was also true. Nixon's own actions as the election approached would have confirmed Kennedy's charge.

But Kennedy didn't run for president in 1972, didn't press his case against Nixon day after day, didn't make the story he had to tell part of the broad national debate.

McGovern almost put Kennedy's story into play. He replaced Sen. Thomas Eagleton of Missouri, the running mate he dropped, with Sargent Shriver, the husband of Ted Kennedy's sister Eunice. Often an afterthought in histories of the 1972 campaign and the Kennedy family, Shriver was the first director of the Peace Corps, the founder of the Job Corps and Head Start. He was not McGovern's first choice—or second or third. McGovern had wanted a real Kennedy, Ted himself, as his running mate; failing that, he settled for Shriver, who was at least Kennedy-adjacent.[3] In the final days of the campaign, Shriver picked up on the story his brother-in-law had begun to tell and provided a glimpse of what might have been.

"No One Will Give a Damn"

As things were coming apart all around McGovern, they came together for Nixon. The first indication that the North Vietnamese were ready to deal appeared on the snack tray at the August 1, 1972, negotiating session. The spring rolls the North Vietnamese served were somewhat thicker than usual. They served cookies, too. Kissinger saw these as auspicious signs. As Zhou had urged, Le Duc Tho now proposed that the *current* Saigon government participate in a three-party coalition government, which he downgraded to provisional. It was still not the powerless three-party election commission that Nixon and Kissinger wanted, but it was a step in that direction.

At Nixon's first meeting with Kissinger afterward, the president got right down to politics. "We've got about 97 days left until the election," Nixon said on August 2 in the Oval Office. They might have to put out the record of the negotiations.

President Nixon: It'll have to get across that the aid and comfort that these clowns have been giving to these people has destroyed a very, very serious effort which we've been making at very great length to settle this thing. I hope the record doesn't have to come out. [*Unclear*] but I meant—that is—that should be our thrust.

Kissinger: I agree completely, Mr. President.

President Nixon: [*speaking over Kissinger*] I [*unclear*], I wouldn't be too damn responsible if I were you.[1]

It was a time for hope. "If you stay ten points ahead, I would say now the chances are two out of three that they'll settle in October," said Kissinger.

"The real question," Nixon said, "is whether, whether we settle at a cost of destroying the South Vietnamese."

"Well, we cannot accept this proposal," said Kissinger.

"Frankly, I'd like to trick them. I'd like to do it in a way that we make a settlement and then screw them in the implementation, to be quite candid," said Nixon.

"Well, that we can do, too," said Kissinger.

The president was talking about a short-term stratagem: "promise something, and then right after the election, say that Thieu wouldn't do it." (Hanoi, as we shall see, wouldn't give him the chance to try this.) "The advantage, Henry, of trying to settle now, even if you're ten points ahead, is that that will ensure a hell of a landslide," said Nixon.

Kissinger wondered whether there was a way to maneuver "so that it can look like a settlement by Election Day" but still leave the final outcome open. "If we can get that done, then we can screw them after Election Day if necessary," said Kissinger. "And I think this could finish the destruction of McGovern."

"Oh, yes. And the doves," said Nixon. "Which is just as important."

"I think we have two problems," said Kissinger. "It isn't just that you win, which is crucial—"

"Got to win big," said the president.[2]

Hanoi's latest concession fulfilled one of Nixon's bottom-line requirements for a settlement. The North had backed off from its demand that Thieu resign before it began negotiations with the South Vietnamese government.[3] Nixon had always told the Communists that he wouldn't overthrow Thieu himself. Once Hanoi dropped the demand for his resignation, a "decent interval" deal came within Nixon and Kissinger's reach. When the president and adviser met the next day in the Oval Office, they had their single most revealing conversation on tape about the "decent interval" exit strategy.

"Now let's look at that just a moment again, think about it some more," said Nixon on August 3, "but . . . let's be perfectly . . . cold-blooded about it. If you look at it from the standpoint of our game . . . with the Soviets and the Chinese, from the standpoint of running this country . . . I think we could take, in my view, almost anything, frankly, that we can force on Thieu. Almost anything. I just come down to that." At that point, the North Vietnamese were badly hurt, Nixon said, and the South Vietnamese were "probably going to do fairly well." Doing fairly well, however, was not the same as surviving after a settlement that would remove all American troops from the South. "I look at the tide of history out there, South Vietnam probably can never even survive anyway. I'm just being perfectly candid," said Nixon. "We also have to realize, Henry, that winning an election is terribly important. It's terribly important this year, but can we have a viable foreign policy if a year from now or two years from now, North Vietnam gobbles up South Vietnam? That's the real question."

Kissinger: If a year or two years from now North Vietnam gobbles up South Vietnam, we can have a viable foreign policy if it looks as if it's the result of South Vietnamese incompetence. If we now sell out in such a way that, say, within a three- to four-month period, we have pushed President Thieu over the brink—we ourselves—I think, there is going to be—even the Chinese won't like that. I mean, they'll pay verbal—verbally, they'll like it—

President Nixon: But it'll worry them.

Kissinger: But it will worry everybody. And domestically in the long run it won't help us all that much because our opponents will say we should've done it three years ago.

President Nixon: I know.

Kissinger: So we've got to find some formula that holds the thing together a year or two, after which—after a year, Mr. President, Vietnam will be a backwater. If we settle it, say, this October, by January '74 no one will give a damn.[4]

Kissinger offered two rationales for a "decent interval" strategy, one foreign, one domestic, both based on credibility.

One rationale was a lot stronger than the other. The notion that China would somehow mind if Saigon fell too *soon* depended on studiously ignoring Zhou's repeated calls for Nixon to withdraw even faster. The geopolitical rationale for a "decent interval" sounds like a rationalization.

The political rationale, however, is airtight. If Saigon fell three or four months after Nixon brought the last troops home, everyone would know he lost the war. Not only Moscow and Beijing, but New York and Peoria and Duluth. Americans would see that all they got for adding four more years to the war was four more years of the Nixon administration. If the president had withdrawn three years earlier, they would have been spared both. The war would still have been lost, but at a far lower cost.

"Idealism with Integrity"

Gov. Ronald Reagan welcomed delegates to the 1972 Republican convention with the perfect words to kick off the president's reelection campaign: "The last American combat team is on its way home from Vietnam." The delegates roared. It seemed like vindication at last. A wave of good news was cresting at the best possible time. It was beginning to look a lot like peace with honor.

Republicans gathered in the same Miami Beach Convention Center that the Democrats had filled one month earlier, offering home viewers a clear contrast.[1] This time things went according to schedule, for maximum political benefit. Not knowing that he was speaking of a president who had timed American military withdrawal to his reelection campaign, who did not believe that prolonging the war for four years had made the South capable of standing on its own, who had secretly assured the Communists they could overthrow the Saigon government without fear of American intervention if they just waited a year or two, Reagan praised Nixon's "idealism with integrity: [Nixon] inherited a full-blown war that had gone on seven years under two presidents. Never once after taking office did he criticize his predecessors for their conduct of the war, nor has he ever charged one of them with killing young Americans for personal political ambition." Reagan made idealism his theme. "President Nixon's idealism is such that he believes the people of South Vietnam should have the opportunity to live under whatever form of government and with whatever society they themselves choose," said Reagan. "President Nixon's idealism caused him to say this nation will do whatever has to be done so long as one young American remains in enemy hands."[2] It sounded beautiful.

Nixon accepted the nomination in a speech that used the word "peace" thirty-six times: "Peace is too important for partisanship. . . . There is no such thing as a retreat to peace. . . . We have the opportunity in our time to be the peacemakers of the world, because the world trusts and respects us." He recalled his pledge four years earlier in that hall "to seek an honorable end to the war in Vietnam. We have made great progress toward that end." As an example of that progress, he declared, "We have reduced our casualties by 98 percent"—the very best face he could put on the deaths of 19,118 Americans in Vietnam during his presidency. (The number of casualties *per month* may have declined, but no one can reduce *casualties* without first acquiring the power to raise the dead. Forty-one more American soldiers would die that month.)

To close, the president invoked the dearest dream of humankind: "I ask you, my fellow Americans, to join our new majority not just in the cause of winning an election, but in achieving a hope that mankind has had since the beginning of civilization. Let us build a peace that our children and all the children of the world can enjoy for generations to come."[3]

In Paris on September 26–27, Le Duc Tho continued to make concessions. Hanoi's overriding goal in the negotiations was the complete withdrawal of all American military forces—ground, naval, and air.[4] The North Vietnamese hoped that the upcoming American election would put pressure on Nixon to settle.

Kissinger told them that with Nixon far ahead of McGovern in the polls, "the only danger we face in the election is not from our opposition; the only danger we face in the election is if we are accused of betraying our allies."

The North's proposal for a three-party coalition government was starting to look more and more like Nixon and Kissinger's three-party election commission. Both could act only by unanimity, meaning neither would act at all.

Kissinger had come up with another synonym for "decent interval," one that confused Le Duc Tho. "We don't understand why you propose that the timing for the reunification will be decided upon 'after a suitable interval following the signing of an overall agreement,'" said Tho. "I don't understand the reason why. I think that this formulation of yours is vague and not necessary."

Kissinger, as he generally did when discussing what he called the "political" issue during negotiations, spoke the language of elections: "We believe that a Committee of National Reconciliation representing three forces should be organized to insure genuinely fair and free elections, so that the Vietnamese could choose their definitive government."[5]

The three-party election commission was just a smoke screen, as Kissinger told Nixon back in Washington on September 29. "You see, Mr. President, this is all baloney, because the practical consequence of our proposal, as of their proposal, is a ceasefire. There'll never be elections. The elections would be run by a committee—or, in their case—by a government of national concord which makes decisions by unanimity. There'll never be an electoral law. They'll never agree on an electoral law on the basis of unanimity," said Kissinger. "Therefore, there'll never be elections. In either case."

"So then what happens? They—they just resume the war later, huh?" asked the president. "But we'll be gone."

"Yeah. This is their face-saving way . . ." Kissinger didn't finish the thought. Nixon and Kissinger's face was being saved by the three-party election commission. It helped with the illusion that who would rule South Vietnam was a "political" question when, as this conversation confirms, it was, in fact, the decisive military question. As Kissinger said, there would never be elections.

The closer Nixon and Kissinger got to agreement with the North, the more they had to worry about confrontation with the South. Nixon and Kissinger were not the only ones who thought South Vietnam wouldn't survive without American ground troops. So did South Vietnamese president Nguyen Van Thieu.

"He doesn't want us out. I mean, let's face it," said Kissinger. "The real point is that our interests and his are now divergent. We want out and we want our prisoners. We want a ceasefire. He wants us in, he thinks he's winning, and he wants us to continue bombing."[6]

"For another two or three years," said Nixon.

"For as long as needed," said Kissinger.[7]

With a deal seeming closer than ever, there was a subtle change in the way the president talked about the war. Before meeting with Soviet foreign minister Andrei Gromyko at the White House, Nixon spoke of drawing a line in the sand. "I intend to hit Gromyko hard," he said on September 30. "They've got to hear that things are going to be goddamn tough." The next round of negotiations would begin in one week, and Nixon was making his final offer: "If they want to settle it diplomatically, this is it." Otherwise, "we're just going to have to seek a military solution, and we will."

"We can do anything that doesn't collapse the non-communist side," Kissinger said.

"The non-communist government of South Vietnam has got to survive."

"Right."

"Period."[8]

Survive, period? That wasn't a "decent interval." Did Nixon mean that? Kissinger needed to know before he closed the deal.

On October 2, the tough talk surged. The president said he might continue to bomb and mine the North for six months *after* the election, even if public opinion turned against it. "So be it. Let it turn," said Nixon. "I have determined that I am not going to sit here and preside over 55,000 American dead for a defeat. Now goddamn it, we're not going to do it."[9]

What did he mean?

To prepare Thieu for what was to come, Nixon and Kissinger dispatched Haig to Saigon. From there Haig reported on "a major confrontation" with just about every foreign policy official in the South Vietnamese government. As he and Ambassador Ellsworth Bunker "entered the palace there was a clear atmosphere of crisis." Instead of seeing Thieu in his office, the Americans were ushered into a large conference room where he had gathered his entire National Security Council. Thieu started assailing the American proposal for a three-party commission (he called it a government) operating by unanimity.

> President Thieu pointed out that under this concept a stalemate would result and culminate in his removal and in the ultimate collapse of the government of South Vietnam. . . . After a lengthy counter-attack by me, the session became highly charged and emotional, with Thieu in tears through much of its duration. I used just about every argument conceivable in an effort to move Thieu and his associates. In the final analysis, Thieu stated that he could not serve as president of South Vietnam and acquiesce in accepting U.S. proposals which would only result in the collapse of his government and contribute to the realization of Hanoi's objectives.[10]

That was just Haig's preliminary report. He later sent a transcript:

> President Thieu: In the proposal you have suggested, our government will continue to exist. But it is only an agonizing solution and sooner or later the government will crumble and Nguyen Van Thieu will have to commit suicide somewhere along the line.[11]

"Our Terms Will Eventually Destroy Him"

October 6, 1972, 9:30 a.m., the Oval Office. The national security adviser needed a decision. In twenty-four hours he would be on a plane to Paris. Kissinger thought Hanoi was ready to settle on the terms that he and Nixon had been seeking for years. The question now was whether Nixon was willing to settle. Kissinger had to know.

Kissinger: I read Haig's transcript, and these guys are scared. And they're desperate. And they know what's coming. And Thieu says that, sure, this—these proposals keep him going, but somewhere down the road he'll have no choice except to commit suicide. And he's probably right. I mean, we—
President Nixon: Let's talk among ourselves [*unclear*].
Kissinger: We have to be honest—
President Nixon: Right.
Kissinger: —among ourselves.[1]

The president wanted to talk about his opponent's latest attack. In a speech outlining his vision of American foreign policy, McGovern "returned repeatedly to Vietnam as an example of what he described as the moral bankruptcy of the Nixon administration," the *New York Times* reported.[2]

"All this is about is morality right now," said the president.

"What's our crime?" asked Kissinger. "That we don't want to destroy—"

"[*Unclear*] that is the morality. We don't want him to—him personally or the 17 million South Vietnamese collectively—to commit suicide!"

"That's right."

"Or to be murdered. Now, that's all this thing is about."

"That is true."

"And goddamn, if that isn't morality . . . ," said the president.

The adviser gingerly began circling back from what they wanted to what they were going to get. "Everything that we ever planned for is happening. The Russians are pressing them. The Chinese are pressing them," said Kissinger. "VIP planes going back and forth between Peking, Moscow, and Hanoi." Le Duc Tho was peppering him with messages. "And I actually think we can settle it. On terms, however."

"On our terms," said the president, "but not Thieu's."

"On close to our terms," said Kissinger. "And I also think that Thieu is right, that our terms will eventually destroy him."

"You're convinced of that, Henry?" asked Nixon, as if they'd never discussed it.

"Not that they shouldn't," said Kissinger. "But given their weakness, their disunity, it will have that consequence."

"And their fear. Fear. Fear," said Nixon.

"And it will be the consequence," Kissinger repeated. No one could say he hadn't warned the president that their settlement terms would destroy South Vietnam.

Kissinger took aim at the notion of continuing to bomb and mine the North for another six months after the election. "Now, the thing that makes it anguishing is that, supposing we don't settle," said Kissinger, "I don't see that we're better off six months from now."

The president didn't accept that. "The chances are better than even that we will not be better off [in] six months, but there is a chance that we could be better off. There is a chance," said Nixon. "It might be effective. We've never done anything militarily that's worth a shit in North Vietnam, except the mining."

"Yeah, but we won't do it this time, either," said Kissinger.

"Let's see. Let's see what we'll do."

"I've had it studied," said Kissinger. Six more months of fighting would improve South Vietnam's situation, but North Vietnam would still hold American prisoners of war. Should the bombing become too much for the North Vietnamese—and it hadn't so far—they could stop it by offering a simple trade: release of the POWs in return for American withdrawal. "I think at this point, we have to take that," said Kissinger.

"We will. I'd take it today," said Nixon.

"Well, I don't think *that* we can do before the election," said Kissinger.

The president agreed. "I mean, I'd take it in November, December, January," said Nixon. "That's a deal we have to take, Henry."

"That's right, but that will also collapse the South Vietnamese," said Kissinger. Without a settlement, there would be no "decent interval." Nixon would bring the troops and POWs home while the fighting between North and South continued—until Saigon collapsed. In that case, Kissinger said, as if to soften the blow, "we won't be so responsible for the whole settlement." There wouldn't *be* a settlement. Just a retreat followed by quick defeat.

"Well, if they're that collapsible, maybe they just have to be collapsed," said the president. "We cannot keep this child sucking at the tit when the child is four years old."

Did Nixon truly mean that? The national security adviser tried a different tack. "See, what we can get out of a settlement now—I'm not even sure it's going to help you politically. You can judge better whether you will wind up like Churchill," said Kissinger. After Winston Churchill led Great Britain to victory against Nazi Germany, voters tossed him out of office in the first postwar election.

"I don't want it before the election, Henry," said the president.

Kissinger didn't point out that, less than twenty-four hours earlier in this very office, the president had told a news conference that "under no circumstances will the timing of a settlement" at all "be affected by the fact that there is going to be an election November 7."[3] It was front-page news. "Election, Peace Bid Separated," said the *Washington Post*.[4] Over a long public career Nixon had developed total immunity to irony.

"Well, if we keep going, you may have no choice," said Kissinger. "You may get it before the election." Hanoi was about to meet his demands. Nixon could hardly reject a settlement on his own terms.[5] That would appear to confirm McGovern's charge that he wasn't willing to end the war.

"Well, let's try our best not to have it before the election," said Nixon. "The more that we can stagger past the election, the better."

"You do *not* want it before the election," said Kissinger.

"Well, I don't want it before the election with a Thieu blowup," said Nixon. If Thieu said in public what he was saying in private—that the settlement terms would destroy the government that more than fifty thousand Americans had died defending—then, as Nixon said, "it's going to hurt us very badly."

"Well, we may be able to avoid a Thieu blowup," said Kissinger. "I've taken the liberty of sending Thieu two letters from you in each of these last two days." In one "personal message from the president," Kissinger told Thieu the opposite of what he'd just told Nixon about Saigon's survival under a settlement:

> There is no doubt that there are serious disagreements between us, but it should be clearly understood that these disagreements are tactical in character and involve no basic difference as to the objectives we both seek—the preservation of a non-Communist structure in South Vietnam which we have so patiently built together and which your heroic leadership has preserved against the most difficult of trials.[6]

The message included a veiled threat: "In this context, I would urge you to take every measure to avoid the development of an atmosphere which could lead to events similar to those which we abhorred in 1963 and which I personally opposed so vehemently in 1968." In 1963, President Kennedy had given the green light to a plot by South Vietnamese generals to overthrow a previous president of South Vietnam, Ngo Dinh Diem.[7] In 1968, there were unfounded rumors of an American coup plot against Thieu himself.[8]

In truth, Nixon and Kissinger couldn't afford a coup in South Vietnam. Not that they hadn't considered it. The question was so delicate that the one administration document to address it directly remained unsigned and named no agency of origin. The NSC considered the analysis important enough to stamp it "Secret Sensitive" and keep it with a set of negotiating records earmarked "For the President's Files." According to that document:

> The U.S. certainly holds the power to remove Thieu or force his resignation and could probably bring this about in fairly short order merely by stating publicly that the U.S. no longer desired to have Thieu as president, or by stating the same thing privately to Thieu and half a dozen other South Vietnamese political and military leaders. At that point, Thieu would lose the support of other key South Vietnamese military and political leaders, and his departure from the scene would only be a matter of time. With Thieu having been removed in such a fashion, however, it would almost certainly be impossible to prevent the rest of South Vietnam's governmental structure from falling apart.[9]

Bottom line: No Thieu, no decent interval. Nixon and Kissinger had to live with him, because without him the South would die too soon.

Kissinger had also instructed Ambassador Bunker to deliver an unveiled threat. If Saigon broke its silence, America would go it alone:

> Thieu must understand that a public confrontation with us would lead to complete disaster. Our only option in that event would be a unilateral disengagement. Please recall to Thieu that our concern is not with the effects on the election but rather with building a platform from which we can take the kind of action we want in the post-election period. If there is [a] public confrontation with [us] now, it will make it absolutely impossible [ev]en to maintain the present level of our military action after the election, much less to step it up. It is almost incomprehensible to the president how the GVN [Government of South Vietnam], after all the risks he has run, could behave so insensitively.[10]

It wasn't so incomprehensible. Kissinger agreed that Thieu was right: their settlement terms would destroy the South. He and Nixon simply couldn't afford to let Thieu say so in public. So he had Ambassador Bunker make the argument that merely revealing this view (that Kissinger secretly shared) would lose the war: "Both the future of South Vietnam and a viable US foreign policy in the world are at stake here. The movement toward a confrontation between us must end if we are not to throw away ten years of effort and the lives of thousands which have been devoted to securing the future we have both sought."[11]

Kissinger hoped to conciliate the South with "security guarantees." One addressed the Ho Chi Minh Trail. For more than a decade Hanoi had been using the border areas of Laos and Cambodia to spirit soldiers and supplies into South Vietnam. Publicly, Nixon stressed the necessity of securing the border, arguing that "we could never succeed as long as we did not seal off the infiltration routes through Laos and Cambodia."[12] His first major decision on the war as president was to start secretly bombing the infiltration routes in Cambodia with B-52s. In 1970, he ordered American troops into Cambodia "to clean out major enemy sanctuaries."[13] In 1971, he backed a South Vietnamese ground offensive in Laos and Cambodia with American airpower. These surges produced short-term results. All disrupted the flow of men and material into the South. None came close to stopping it. Nixon would claim to have "cut" the Ho Chi Minh Trail, but that was a metaphor without real meaning.[14] The border was a thousand miles along. How do you cut that? Not even B-52s could level an area that vast. Ground offensives captured more supplies and killed more soldiers—until the offensives ended, as all must, and American and South Vietnamese troops withdrew back across the border. Then the infiltration resumed. Hanoi managed to replace its losses in a matter of months.

At the negotiating table, Kissinger would seek "a commitment by the North Vietnamese to withdraw from Laos and Cambodia."

President Nixon: I'd get the commitment. I wouldn't worry about it. They're never going to withdraw.
Kissinger: No, but I'll get it in writing and we'll—
President Nixon: Right, right, right.[15]

So much for cutting the Ho Chi Minh Trail. With the paper commitment in hand, Nixon would at least be able to claim he'd finally achieved the goal of closing it. In reality, it would remain open for business, as he privately acknowledged.

Another "security guarantee" would address the heart of the matter: the continued presence of North Vietnamese troops in South Vietnam. The threat that, according to the consensus of military, diplomatic, and intelligence advisers in NSSM-1, Saigon couldn't withstand without American ground troops. Kissinger would ask Hanoi to bring some of its troops home: "They withdraw from—some of their units from the South."

"Some of their units, right," said the president.
"They won't withdraw them all," said Kissinger.

Kissinger's security guarantees all had one thing in common: they wouldn't guarantee Saigon's security.

The national security adviser had one more card to play: "As I look at it from a historical point of view, what did [French president Charles] de Gaulle do in Algeria, who everyone thinks a great man? Basically, he made a settlement that turned the country over to the—to his enemies." De Gaulle had ended Algeria's war for independence by putting it up for a vote. No surprise who won. Algerians outnumbered French *colons* and voted, as they had fought, to end colonial rule. With de Gaulle's backing, France in referendum also endorsed independence.

It wasn't the best analogy. Unlike Algeria, South Vietnam wouldn't get to vote on its fate. Its electoral commission, requiring unanimity to do anything, was designed to deadlock. North and South would fight it out after American troops left, and, as Kissinger had pointed out twice in this conversation, the North would win. Americans would not get to vote on this "decent interval" exit strategy; the White House kept its very existence a secret, all records of it highly classified.

But Kissinger wasn't appealing to logic. He was invoking the name of a man Nixon viewed as a role model, almost a hero. Charles André Joseph Marie de Gaulle (Nixon savored his full name, redolent of ancient Gaul and Charlemagne) dazzled the American president as "a larger-than-life figure," "a living legend," and "a master of illusion." De Gaulle asserted control over his image to a degree Nixon could only envy, scripting his press conferences (not just the answers, but the questions his aides planted), memorizing speeches so France would not see he needed glasses to read, and generally exuding grandeur.[16] Nixon would never forget the splendid climax de Gaulle provided to his first trip abroad as president: the honor guard and red-carpet welcome when Air Force One touched down at Orly Airport.[17] The roads of Paris were completely cleared in both directions to make way for the motorcade sweeping the Americans off to the Quai d'Orsay Palace.[18] Dinner at the Élysée Palace, lunch at Versailles. Above all, Nixon prized the time he spent alone with de Gaulle, enthralled by the older man's political wisdom. "I make policies for the newspapers of the day after tomorrow," de Gaulle told him. It would be better for the United States to recognize China now, when it didn't have to, than later, when Chinese growth would make recognition a necessity, de Gaulle advised. And he urged Nixon to get out of Vietnam as quickly as possible, though not "with undue haste." He held up Algeria as a precedent.[19] Three years later in Beijing, Zhou Enlai had also urged him to follow de Gaulle's Algerian example.[20]

Nixon would always remember de Gaulle's private parting words in March 1969, when the French president broke protocol to personally escort the American to the airport. As the limousine carried them through the mists of Paris, de Gaulle turned to Nixon, took his hand, and said, "You look young and vigorous and in command. This is important. Stay that way."[21]

"Let me gallivant around in a different way," Nixon told Kissinger. "Vietnam is important because, of course, of our prisoners and, of course, because we don't want 17 million people to come under Communists, but . . . However, those, basically, are not the *really* important issues. The important issue is how the United States comes out in two ways.

"One: whether or not the United States in all parts of the world—whether our enemies, the neutrals, and our allies after we finish—are convinced that the United States went the extra mile in standing by its friends. That doesn't mean we have to succeed. It does mean that we have to have done that.

"Second point: Now, the historical process moves extremely slowly," Nixon said. A lot could happen within the South Vietnamese government. Someone could retire or be exiled. There could be a coup. "I think the North Vietnamese are hurting one hell of a lot more than the CIA indicates. I think our whole bureaucracy previously overestimated how badly they were hurting and now they're underestimating how badly they're hurting, because we're doing it. The mining has had to hurt them. The bombing has had to hurt them. It's supposed to be pretty good. It's just got to have done it," said Nixon. "Now, they're hurting, and hurting badly."

President Nixon: Let us suppose, putting it quite coldly, that we face a situation here where South Vietnam simply, over the long haul, cannot survive on its own as an independent entity. I don't mean like Thailand, where we can get a guarantee. I don't mean like [*unclear*]. But that South Vietnam, because of its—the nature of the South Vietnamese people, the nature of the struggle with the North and so forth, that inevitably, unless the United States can stay in there indefinitely, South Vietnam is going to fall.

All right, if that is the case, then what we have to look to is the bigger subject: How does the United States look in the way it handles this goddamn thing? So as I see it, the thing to do is to look as well as we can and hope and pray for the best. And then use our influence with the Russians and with the Chinese, which should be considerable at this point, and say, "Now, damn it, you push us here, you know, we contribute—we just cannot be pushed too far." Understand?

Kissinger: Yes.[22]

If he had wanted, Kissinger could have demolished the president's reasoning. He knew the way the United States looked to Moscow and Beijing. Its government had long ago informed the Soviets and the Chinese that North Vietnam could take over the South without fear of American intervention as long as it waited a "decent interval." This was a secret from America's citizens, but not from its most powerful adversaries.

Kissinger, however, wasn't there to win an argument. He'd gotten what he came for: the go-ahead to make the deal.

"I almost think that that's what we're looking at," said Nixon. Having passed the point of no return, Nixon exercised the presidential prerogative of hedging a little: "My own view is this, that when Defense is planning for three more years of bombing, when they want to keep two more carriers out there, three more air wings and all that sort of thing, it makes me think that the military isn't particularly interested in finishing this goddamn war." The military, of course, wasn't going to decide when America's war would end; that decision rested in the hands of the commander in chief.

Nixon's moment of truth had come and gone. At least he wasn't talking about the need for Saigon to survive, "period." Kissinger egged him on: "The military are a bunch of selfish bastards."

"They screwed up everything we've done," said the president, "except they did the mining OK."

One final detail: Hanoi wanted the agreement to bear the signatures of all parties' foreign ministers. That meant Secretary of State William Rogers would have to sign it for the United States.

"It's not fair—let him sign it—but we don't care," said Nixon.

"Oh, I don't want to sign the goddamn thing," said Kissinger. "You should sign it."

The president declined. "I don't think we should dignify it by my signing it."[23]

Blowup 1968

Nixon understood better than anyone else the dangers of a public outburst by the president of South Vietnam. A Thieu blowup in 1968 had practically decided the outcome of the presidential election. Nixon had started out with a huge, 16-point lead over Humphrey in the polls. But in October, news reports of a peace initiative in the works narrowed the gap. On October 31, less than a week before the election, President Johnson went on television and announced his decision to halt the bombing of North Vietnam. The North, which had refused to talk to representatives of the Saigon "puppet" regime, agreed to sit down with representatives of the South as well as the United States in Paris. Humphrey, as LBJ's vice president, profited by association with the peace initiative. Nixon's lead in the polls evaporated. In Gallup the two were within the margin of error; in Harris, Humphrey pulled ahead. But on November 2, the Saturday before the election, Thieu announced that Saigon would boycott the Paris talks. "This would be another step toward a coalition government with the Communists in South Vietnam," said Thieu. Coalition was a dirty word in South Vietnamese and American politics, viewed as a synonym for defeat. Thieu's blowup made the peace initiative look like a slimy political trick designed by LBJ to boost Humphrey. The Democratic campaign's rise stalled. Nixon won, but it was too close— the second-closest presidential election in the twentieth century. Only Nixon's loss to JFK had been closer. The president would do a lot to avoid a blowup by Thieu.[1]

"We're behind the Trees!"

October 8, 1972, 10:30 a.m., Gif-sur-Yvette, France. A pace-setting anecdote, courtesy of Henry Kissinger: "I am told there's one race track in Paris, in Auteuil, where when they get around the other side—they're behind the trees so you can't see them—and I'm told that that's where the jockeys decide who will win."

Le Duc Tho asked, "Are we making now a race to peace or to war?"

"To peace," Kissinger said, "and we're behind the trees!"

Hanoi's chief negotiator came bearing two big green folders. Kissinger wanted to know what was in them, but the North Vietnamese made him wait. They asked him to speak first, to respond to their previous proposal, to offer anything new he might have. This took over an hour. Le Duc Tho suggested they break for lunch. The snack assortment seemed generous—a good sign.

As they ate, Kissinger tried to sell Le Duc Tho on some sort of security guarantee regarding the Ho Chi Minh Trail, even though he and Nixon knew infiltration would continue.

"When the war ends, some day I will show you the Ho Chi Minh Trail!" joked Le Duc Tho.

"Seriously," said Kissinger, "we cannot go to Saigon on the political proposal without the security provisions."

Le Duc Tho left for a moment to talk with his delegation and returned to ask for three more hours to digest all the information Kissinger had given them. The Americans stepped out into the cool, crisp autumn air, whispering among themselves for fear that the garden was bugged. The French Communist Party owned the spacious white villa, a gift from a Cubist whose abstract paintings and tapestries provided the backdrop for their deliberations.

At 4:00 p.m., Le Duc Tho opened a green folder. The North had a new proposal. It included Nixon's essentials: release of the POWs, total withdrawal of American forces, a ceasefire-in-place for the North and South Vietnamese. Hanoi had dropped its demand for a coalition government. Hanoi proposed American reparations, without using that politically unacceptable word, referring instead to "the question of U.S. responsibility to heal the war wounds and to rehabilitate the economy of Vietnam." For the tripartite body that would arrange South Vietnamese elections (in the unlikely event that its three parts could ever achieve unanimity) Hanoi was now proposing an "administration of national concord," which still sounded too much like a coalition government for it to fly politically in either Saigon or the United States. President Thieu would not have to resign upon signing, but he would have to step down before the elections (if, against all odds, they were to actually take place).

Other parts of the proposal were not so pretty. Hanoi insisted on "the principle of equality" in replacement of armaments. If the United States could resupply Saigon with arms and ammunition, Hanoi would do so for its troops in the South. This made it all too apparent that what the two sides were planning behind the trees was only interrupted and to-be-resumed war, not peace. It also rendered slightly ridiculous the provision in Hanoi's proposal to "strictly respect" the sovereignty of Laos and Cambodia, since infiltration via the Ho Chi Minh Trail through both countries was *how* it would resupply its troops.

By and large, however, Le Duc Tho displayed exquisite sensitivity to the Nixon administration's political requirements: "When we put forward this new proposal we do not let the political problem of South Vietnam, that is the most thorny, the most difficult problem, to drag out, to prolong our negotiations; and we should aim at rapidly ending the war, responding to the aspiration for peace of our two peoples. At the same time we have taken into account the questions on which you have shown the greatest concern. Last time, Mr. Special Advisor said that there was a danger, the greatest danger for you in the U.S. election, this danger comes from the part of your supporters who would denounce you to have betrayed your ally."[1]

After nine hours of negotiating, Kissinger emerged elated.[2] He was so close.

His team was not entirely sold. John D. Negroponte ("my Vietnam expert," as Kissinger called him) expected Saigon to react angrily. Kissinger had Negroponte and Winston Lord draft a counterproposal, but when he saw the results, he exploded, wrote Seymour Hersh in *The Price of Power:* "'You don't understand,' Kissinger shouted, according to an eyewitness. 'I want to meet their terms. I want to reach an agreement. I want to end this war before the elections.' Negroponte got upset in return and accused Kissinger of 'trying to do too much with too much risk.'"[3]

Negroponte later said that for Kissinger, "the scenario was almost more important than the words" of the agreement: "He got so excited about going to Saigon and then going to Hanoi to announce signing."[4]

Kissinger gave them until noon the next day to come up with something more accommodating.[5]

Back at the villa, Kissinger apologized twice in rapid order for keeping Le Duc Tho waiting. He accepted the basic approach and many of the provisions Le Duc Tho proposed, but suggested changes to make it politically acceptable in Saigon and the United States. "And our opposition will come from the right," said Kissinger, "not from the left."

He returned once more to his need for a security guarantee regarding the Ho Chi Minh Trail. "Now I know," Kissinger said, "speaking frankly, and the Special Advisor knows, that if you are determined to move supplies through Laos you will find a way of doing so. You always have. Or am I wrong?"

Le Duc Tho laughed.

"On the other hand, we rely on the fact that you will consider this inconsistent with our long-term relationship and that therefore you will look at problems henceforth in a different way," Kissinger said. "Yet to increase the acceptability of the agreement in the United States and to speed up the deliberations in Saigon, if we in the next day or two could find some formula to make this possible, it would be very important." Kissinger cautioned the North Vietnamese to avoid public hints that they'd made any progress, since he was planning not to show the agreement to Thieu until he arrived in Saigon. At the end of the day's negotiations, Kissinger quipped about infiltration: "When the Special Advisor goes to bed tonight and he is thinking about Ho Chi Minh Trail, maybe some ideas will come to him." This time Kissinger laughed as well.[6]

"Saving Face or Saving Lives"

What better moment for the challenger to take to the television airwaves and eviscerate the incumbent's strategy for Vietnam, right as the national security adviser was selling out Saigon for a "decent interval." McGovern bought a half hour of airtime on the CBS television network on October 10, 1972, to give the most important speech he would ever make on Vietnam, the defining statement on the defining issue of his campaign, his career, his life. It was the perfect time to turn the tables on Nixon and say that it was the president who was negotiating "peace with surrender," who had sacrificed the lives of thousands of American soldiers to avoid sacrificing his chance of reelection, who would reap nothing from four bitter, divisive years of war except a deal with the Communists for delayed, disguised defeat.

McGovern didn't say that.

Seated in an armchair in a corner of the Capitol office of the Senate leader, McGovern looked into the camera and defined the election this way: "It is a choice, after all, between saving face or saving lives. It is a choice between four more years of war, or four years of peace."

If Americans still had any doubts about the Republican's commitment to preserving the Thieu regime, the Democrat put them to rest. "Now, Mr. Nixon would continue the war to preserve General Thieu's power," said McGovern. "Incredible as it seems, when all is said and done, our purpose in Vietnam now comes down to this—our policy-makers want to save face and they want to save the Saigon regime of General Thieu." His faith in Nixon's steadfastness and loyalty was fervent, if misplaced. (And utterly self-defeating. "The more the McGovern side tried to say that the president was in *favor* of the war, the more it worked to our advantage," said Peter Dailey, president of the November Group, the ad hoc ad agency of the Committee for the Re-Election of the President, or CREEP.)[1] In McGovern's defense, it should be noted that he seems to have truly believed every word. The nominee spoke in the prophetic mode, framing America's choice in biblical terms: "I have set before you life and death, blessing and cursing. Therefore, choose life, that thou and thy seed may live."

McGovern made some good points. "Our problem is that we have asked our armed forces to do the impossible—to save a political regime in Saigon that doesn't even have the respect of its own people." If the South Vietnamese defended their regime as fiercely as the North Vietnamese fought for theirs, they would never have needed Americans to do it for them.

But McGovern went too far. The previous decade had proven that Americans *could* save Saigon for as long as they were willing to sacrifice their children to the cause. Doing so was unnecessary but not, as McGovern claimed, "impossible." What no country could do was go back in time and change the history of colonialism that aligned most of Vietnam in opposition to the local functionaries of the French regime who now filled most government positions in South Vietnam.

McGovern went through the standard liberal litany. The war cost too much money that could better be spent back home: "Each week it costs $250 million that we need to rebuild our cities, to fight crime and drugs, to strengthen our schools, and to assist our sick and elderly." Most Americans had already demonstrated their willingness to pay that price, however, when they thought it would prevent defeat. None had demonstrated the willingness to spend that much merely to *delay* defeat. McGovern didn't make the case that the money was wasted on a cause that would end in expensive failure.

He decried civilian casualties: "The reality of this war is seen in the news photo of the little South Vietnamese girl, Kim, fleeing in terror from her bombed-out school. She has torn off her flaming clothes and she is running naked into the lens of that camera. That picture ought to break the heart of every American." Most voting-age Americans had seen the photo. It would win a Pulitzer Prize. It did not, however, lower public approval ratings for Nixon's response to the Easter Offensive. Most Americans accepted unintended civilian casualties as the price of avoiding defeat. It's doubtful that any but the worst extremists would accept such casualties merely to avoid the defeat of a preferred presidential candidate, but McGovern never put that proposition to the test.

His best point was about the POWs: "Now Mr. Nixon says we must bomb and fight to free our prisoners. But just the reverse is true. We must end the bombing—end the fighting—if we're ever to see these prisoners again. Prisoners of war come home when the war ends—not while the war continues."[2] McGovern didn't realize that the president was perfectly willing to end America's Vietnam War and gain release of the POWs now that the risk of Saigon collapsing before Election Day had practically vanished.

Contrary to both liberal and conservative belief, Nixon could accept Communist victory in Vietnam. He just couldn't accept Democratic victory in America.

In public, Nixon would stop short of accusing his opponent of treason. Not in private. "You've got to paint this son of a bitch as for what he is: he's disloyal to this country," Nixon told Haldeman and Chuck Colson.

"He's for imposing a Communist government on the people of South Vietnam, I'm against a Communist government. He's for a Communist government in South Vietnam, I'm against a Communist government."[3] Both McGovern's announced exit strategy and Nixon's secret one would produce the same basic results: complete American withdrawal, release of American POWs, the collapse of South Vietnam, and Communist victory. Nixon differentiated himself from his opponent in private by saying that McGovern had treasonous intent.

Nixon could hate McGovern or hate himself. He could look in the mirror and see a politician who would do anything to win reelection, up to and including delay the day of defeat by prolonging a war and forfeiting American lives for as long as it took to avoid being held accountable on Election Day for losing Vietnam. Or, if he wished, he could see himself as America's secret savior, rescuing the nation from a politician who was somehow worse than he—one who would not merely accept Communist victory in Vietnam, but who secretly desired it in his treacherous heart. The evidence against Nixon was abundant and hard; against McGovern, nonexistent. But the president's men didn't demand evidence. Nor did the president. If he could see evil in his opponent, he could avoid seeing it in himself.

"Brutalize Him"

October 12, 1972, 6:45 p.m., Executive Office of the President. Haldeman had driven all the way home before he got a message summoning him to Nixon's "hideaway" office next door to the White House, where he found the president ensconced in his easy chair with a glass of wine. (Nixon's hideaway office had a bar.) This would be interesting. Haldeman had seen a single beer turn the president's speech slurry. At times he sounded drunk without drinking at all.[1]

The president was reliving the day's campaign trip to Atlanta for the edification of Chuck Colson. "I said, 'Many people deprecate the South because it has religious values.'"

"That's right," said Haldeman.

"And I said, 'But the day that America ceases to have moral and spiritual and religious strength, this country will be finished.' Wow!"

Haldeman had been there. The crowd started applauding quietly, then got louder and louder, and then part of it gave a standing ovation.

"They know this is the great issue with McGovern," said Nixon, "that McGovern is permissive, he's amoral, he's got all these kooks and nuts around him, and I don't have." The men around him did not disagree.

At 7:05, the reason for the meeting walked through the door with a big red folder and a big smile.[2] "See if Dr. Kissinger would like a little drink," the president told his valet.

Scotch and soda for Kissinger. Al Haig, looking less than happy, ordered a martini.[3] "Well, you got three out of three, Mr. President," said Kissinger.

"Do you agree, Al?" asked Nixon. "You're too prejudiced, Henry. You're so prejudiced and peacenik that I can't trust you. Do you think so, Al?"

"Yes, sir," said the general.

"What about Thieu?" Nixon asked.

"Well, that's the problem, but here is the plan," Kissinger said.

Haig characterized the problem with less than boundless optimism as "not insurmountable."

"Not insurmountable? How do we handle him?" asked the president.

"Here is what we have to do," said Kissinger. The national security adviser had to go back to Paris to finish the deal, then to Saigon to sell Thieu, and then to Hanoi to sign it October 31. One week before Election Day. "That was the price we had to pay," said Kissinger.

"Well, that's no price, if we get Thieu aboard," said Nixon. "What do you think, Al?"

More than one man replied; Kissinger won. "The deal we got, Mr. President, is so far better than anything we dreamt of," said Kissinger. "I mean, it will absolutely, totally wipe out McGovern."

"Good."

"The deal is—"

"It won't totally wipe out Thieu, Henry?" asked Nixon.

"No. Oh, no. It's so far better than anything we discussed." (In his memoirs, Kissinger would write more accurately that the North Vietnamese proposal "in effect accepted what we had put forward on May 31, 1971, and on May 8, 1972.")[4] The national security adviser started plowing through the provisions. He got through exactly one (ceasefire-in-place) before Nixon interrupted to note that the French diplomatic mission in Hanoi was hit during an American air raid while Kissinger was in Paris. The Pentagon said a North Vietnamese anti-aircraft missile, rather than an American bomber, may have been to blame. Six people died.[5]

"The silly-ass thing of some SAM hitting the French emb—consulate and everything raises hell about it, didn't make any difference," said the president. "Most people would rather kill all the Frenchmen anyway." The king was in his cups. "My point is, Henry, I'm thinking of Americans. Most Americans are very cynical about all these things now. But the point is that we can't go on and on and on and on having these things hanging over us."

"Sure," said Haig.

"And the other thing," said the president, "is, what're you going to do, nuke 'em?"

Kissinger plowed on: Withdrawal of American troops within sixty days of signing. Continued US military aid to Saigon in the form of replacements.

"Did you and Al hear the goddamn McGovern speech?" Nixon interrupted. "Did that do anything to your negotiations? Did that hurt?"

"It would have if the guy were 15 points closer in the negotiations, Mr. President. It would have killed us," Kissinger said. He resumed plowing. Thieu could stay in office at the time of signing. Regarding elections, there'd be a tripartite "National Council for Reconciliation and Concord" that would function only if the Communists, neutrals, and anti-Communists on it reached unanimous agreement.

"Thieu will never agree, they will never agree, so they screw up and we support Thieu and the Communists support them and they continue fighting, which is fine," the president said. "Now, what did you do with regard to reparations and the rest?" "Reparations" was a politically dangerous word. It implied guilt, amends for past wrongs. Nixon avoided it in public.[6] "I'd give them everything. Because I see those poor North Vietnamese kids burning with napalm, and it burns my heart."

Kissinger called reparations "our best guarantee that they'll observe the agreement." As if money would persuade Hanoi to give up.

"Give 'em ten billion," said the president. (The entire American foreign aid budget was $1.8 billion.)[7]

Kissinger mentioned the part of the agreement dealing with international supervision and control: "It is incomprehensible, but—"

"Keep it that way," the president said. Kissinger mentioned Laos and Cambodia, but the president said, "We don't care about that."

"Let me come down to the nut-cutting," the president said, suddenly very serious. "What Henry has read to me, Thieu cannot turn down. If he does, our problem will be that we have to flush him. And that will flush South Vietnam." No Thieu, no decent interval. The conclusion, to Nixon, was clear: "Basically, you've got to brutalize him," Nixon said. "Thieu is going to accept this. This is a hell of a deal for him. Isn't it, Al? It's a hell of a deal for Thieu?"

"Well, if he's smart," said Haig.[8] (Haldeman checked with the general later to see if he honestly felt the deal was good; Haig said it was.[9] Haig expressed a different view in his memoirs: "It was too late to argue the self-evident truth that Thieu's regime could not survive if we permitted a huge Communist army to remain inside his country.")[10]

The president ordered steaks for them all to celebrate. Haldeman noted that for once the boss also ordered "the good wine, his '57 [Château] Lafite-Rothschild, or whatever it is, to be served to everyone. Usually it's just served to the P and the rest of us have some California Beaulieu Vineyard stuff."[11]

The same day Nixon toasted Kissinger with the good stuff, the president of South Vietnam told a youth rally in Saigon, "We have to kill the Communists to the last man before we have peace." Thieu had counterpointed the secret sessions where Kissinger and Le Duc Tho had worked out his fate with a burst of public appearances where he preemptively rejected unacceptable concessions. "His speeches have followed the theme that South Vietnam will decide for itself what is an acceptable peace settlement," noted the *Washington Star-News*. No coalition government. No neutrality. No Communist activity in the South. No concession of land to the Communists. "We cannot let this happen to South Vietnam," declared Thieu, poking a finger at the sky.[12]

Joe Kraft nailed it. The columnist raised "a new question: whether President Nixon's political interest isn't better served by waiting until after the election before bringing negotiations to a head." The risk to Nixon was that Saigon might object to the deal.

> But given Sen. McGovern's dismal campaign, President Nixon has no special need to get a deal before Election Day. Indeed, it is hard to think of anybody now opposing the President who would be won over by a Vietnam settlement. If anything, on the contrary, the kind of settlement that is bound to emerge would probably cost the President supporters on his right wing.
>
> So the best timing for President Nixon is a settlement made sometime after the election when he is invulnerable to pressures from Gen. Thieu.[13]

Max Frankel of the *New York Times* told Kissinger he was writing a piece for the Sunday paper, saying that "since a settlement was not needed in the bold sense of the word—you know, to win the election—that [one] probably would see some advantage in not setting Thieu screaming." This was pretty close to what Nixon had said ("I don't want it before the election with a Thieu blowup"), but the national security adviser managed to sidestep. "Yeah, but on the other hand we are not quite at that point, but it's a hell of a responsibility to say I'll keep the war going three weeks longer than I have to, because I'll look a little better in tomorrow's newspapers," said Kissinger. "We are not at that point, so I don't want to mislead you, but actually I would have thought from an election point of view it is sufficiently unnecessary for it to be no consideration."[14] Technically, that was not a lie, since Kissinger was giving his opinion rather than the president's.

A lesser standard applied when speaking with Bob Toth of the *Los Angeles Times*. "From our point of view, we are not gearing it at all to the election. It is really astonishing to me how people could believe that it could be geared to the election," Kissinger told the reporter. "Look, Bob, we have suffered four years of anguish on this issue for one reason only—to maintain the dignity and character of America as we saw it. We are not turning this into a tawdry election ploy."[15] (Soon, after Kissinger conceived another tawdry election ploy, he would turn to, and on, Toth.)

At this point, Nixon was willing to settle before the election. He calculated that it wouldn't make much difference politically. "The making of a settlement is not going to hurt us in the election and it isn't going to help us significantly," Nixon told Kissinger. "We don't need it. We're going to win without it. And very heavily."

Nixon could even take a preelection blowup by North Vietnam. Hanoi might try to force Nixon to end the (American) war by revealing that it had met all his demands. "See there, you do run the risk, too, that they might decide to go public and say Thieu is at fault," said Nixon. "However, that's dangerous for them, too, because even with that, we're not going to lose. Haha."[16]

What Nixon would not accept was a Thieu blowup before Americans voted. "I did a lot of thinking last night about Thieu, as to how to set him up," Nixon told Kissinger. "I have a feeling that he ought to say something right now—it's about time that he did—that he has confidence in the president."

"Well, let's wait till I get there," said Kissinger. "We are working him over." Ambassador Ellsworth Bunker was under orders to see the South Vietnamese president every day and soften him up.

A denunciation of the deal by the leader of the land Nixon was supposedly saving could do a great deal of damage. "It won't defeat us, but it could hurt very badly," said Nixon. "Whatever cost needs to be paid, we cannot have any break with Thieu before the election."

"All right."

"Whatever it costs."[17]

The day Kissinger left for Paris, October 16, the president made a campaign appearance before the annual convention of the National League of Families of American Prisoners and Missing in Southeast Asia. Nixon didn't want to go there, but Bob Woodward and Carl Bernstein left him little choice. The *Washington Post*'s investigative duo had published a series of front-page exposés tying the Watergate break-in to "a massive campaign of political spying and espionage conducted on behalf of President Nixon's reelection."[18] This morning's headline: "Lawyer for Nixon Said to Have Used GOP's Spy Fund." Herb Kalmbach, the president's personal attorney and longtime fund-raiser, authorized "payments from the Nixon campaign's secret intelligence gathering and espionage fund," the *Post* reported. At one point the fund totaled $700,000 and financed "more than 50 undercover operatives engaged for more than a year in an apparently unprecedented political spying and sabotage effort staged by Nixon aides against the Democrats," the story said. The undercover campaign, according to federal investigators, included:

> Following members of Democratic candidates' families and assembling dossiers on their personal lives; forging letters and distributing them under the candidates' letterheads; leaking false and manufactured items to the press; throwing campaign schedules into disarray; seizing confidential campaign files, and investigating the lives of dozens of Democratic campaign workers.
>
> In addition, investigators said the activities included planting provocateurs in the ranks of organizations expected to demonstrate at the Republican and Democratic conventions.[19]

The Watergate break-in was the iceberg's tip; Woodward and Bernstein were providing a glimpse of what lay beneath the surface. They'd revealed that one of the undercover operatives had a contact in the White House: Dwight Chapin, the president's appointments secretary.[20]

The White House "had no story to counter the espionage/sabotage story so I went in and talked [Nixon] into going over at the last minute to do the POW wives," Haldeman told his diary.[21]

"They're writing stuff on all this crap," Haldeman told the president.

"Well, I can't go, Bob." He didn't explain why. A year had passed since he last spoke before the POW wives—a year he had kept the war going so Saigon wouldn't collapse before Election Day. A year he added to the POWs' captivity, since Hanoi refused to release American prisoners until Nixon withdrew American troops. How could he face them?

"They couldn't take Henry?" asked Nixon. Kissinger was in the Oval Office with them.

"Yeah, but that won't make news," said Haldeman.

"Are you *kidding?*" asked the president.

Kissinger wouldn't be the top story on the nightly news; Nixon would. "McGovern's turned them down," said Haldeman. "He isn't going to appear at all."

The president fell silent for four seconds. "OK, I'll do it," he said.

"I tend to agree with Bob," said Kissinger. "I think if you gave an emotional three minutes . . . how they're inspiring us, that they're the ones . . ."

Haldeman acknowledged that an appearance by Nixon might draw criticism.

"That I'm exploiting the POW issue," said Nixon.

"Henry's not political," said Haldeman.

"But what would I say?" asked Kissinger.

"Here's what I would say, if I were attending," said Nixon, writing his aide's applause lines while referring to himself in the third person: "'I know that there's nothing closer to the president's heart,' and so forth and so on"; "'The president refuses to exploit POWs for—this issue for political purposes'"; "'The president has had to make hard decisions. For example, his decision on May 8th'" to bomb and mine the North. "'He's had support for some of these decisions and some opposition. But the support that has meant the most to him has been the support of you, whose loved ones are POWs,' and so forth and so on. 'Because he knows what it would mean to you to get them back. But you are insisting that this country must have peace with honor. And that's what we're going to do.' How's that sound to you?"

Kissinger said he couldn't do it later than 10:30; it was past 10:00.

"I think actually, you can see the problem with my doing it," said the president. "I think it looks like a crass attempt to . . . or do you agree or disagree? I'll go over and do the goddamn thing, but [unclear]."

Haldeman left with the answer he wanted.

"Last time, it was a terrific hit," said Kissinger.

"I know," said the president. "A hard year has passed. A year of more suffering and demagoguery and hopes dry . . . just go. You can go to the well once too often." He turned to Paris briefly. "The main thing is, Henry, we're going to—we're going to beat these bastards. We are going to beat them," said the president. "I mean the political opposition are going to be beaten. That's why I say, 'Don't worry about the election.' We're going to beat them."

"I think you won the election on May 8th," said Kissinger.[22]

A thousand people—wives, parents, brothers and sisters, children of the POWs—applauded as Nixon entered the Presidential Ballroom of the Statler-Hilton Hotel. As they caught sight of the president their cheering swelled.

"I learned that Dr. Kissinger was scheduled to be your speaker this morning; I found that I had some time in my schedule and I decided to substitute for two very important reasons: One, of all the many groups I speak to and that I have spoken to, big or small, across America, there isn't one that inspires me more than you do. Two, I am here to thank you for your support and to urge you for your continued support.

"I am not speaking of an election campaign, but I am speaking of support for a cause bigger than an election, a cause of an honorable peace, one that will contribute to peace in the world." The president said he would insist on two conditions: "First, we shall not agree to any settlement which imposes a Communist government upon the people of South Vietnam. And second, we shall, under no circumstances, abandon our POWs and our MIAs, wherever they are." The families cheered the president's insistence on the safe return home of their loved ones.

Nixon raised another winning issue, framing it in moral terms: "It would be the most immoral thing I could think of to give amnesty to draft dodgers and those who deserted the United States. Your loved ones have and are paying a price for their choice, and those who deserted America will pay a price for their choice."

Standing ovation.[23]

Later that evening, the Rev. Billy Graham, the religious and moral leader of millions, got a call from the president. "On all three networks," said Graham, "they carried what you had to say about amnesty—"

"Good."

"—which was tough and right. They showed people in tears at the end of your talk." Graham offered to do anything Nixon wanted before the election. "I'm not in a position to know all that I could do, but you just tell me and I'll do it."

"You don't need any guidance," the president told the preacher. "Your political instincts are very good."

"I'm going to hold a press conference tomorrow afternoon in Tulsa," said Graham, "and I was going to talk about—and I'm so glad to talk to you tonight—about your personal character and your personal morality and integrity and that kind of thing."

McGovern, picking up on the *Post*'s reports, had accused Nixon's administration that day of "the shabbiest undercover operations in the history of American politics."[24]

"I don't think this business is touching you," said Graham, "but you don't want to take any chances."[25]

Kissinger v. Thieu

On the plane waiting at Andrews Air Force Base to take him to Paris, Kissinger found a handwritten note from the president. It was a sanitized version of his oral instructions, minus the political calculus.

> Dear Henry,
> As you leave for Paris I thought it would be useful for you to have some [of the] guidance that we were talking about on paper. First, do what is right without regard to the election. Secondly, we cannot let a chance to end the war honorably slip away.

Kissinger kept the note. He would need it.[1]

Things went almost too well back at the white stucco villa in Gif-sur-Yvette. Twelve hours "resolved all substantive and technical issues," Kissinger reported to Washington on October 17, "except replacement (article 7) and prisoners (article 8)."[2] The North Vietnamese were pressing for the release of prisoners held by Saigon on suspicion of being Vietcong civilian cadres. The number was enormous.

"The difficulty we now face," Kissinger told them, "is that you are first asking that all of your forces can stay in the South. The second is that on top of this you are then asking Saigon to release some 30,000 individuals whom they consider as likely to engage in military activities against Saigon. I have said to the Special Adviser that if I present such a package in Saigon the possibility of our getting agreement in Saigon is very slight, and the probability that we will face a repetition of events in 1968 is very likely."[3] Kissinger wanted to leave the issue for negotiations between the North and South. Hanoi was ready to accept Kissinger's proposal regarding one-for-one replacement of armaments, but only if he agreed on the release of prisoners. Kissinger suggested another round of talks, perhaps after he visited Saigon.[4] (This would have delayed things nicely for Nixon.)

He left Paris without a completed agreement. This made blowups by Hanoi or Saigon less likely since Hanoi couldn't reveal the provisions of a deal it didn't yet have, and Saigon couldn't denounce a deal that wasn't yet done.

Indignity greeted Kissinger's October 19 arrival in Saigon. The city's largest-circulation daily paper had published a picture seeming to show the national security adviser naked and reclining on a Chinese panda-skin rug. The image came from the *Harvard Lampoon,* which had pasted Kissinger's head on the body of a Boston cabbie. The American pop-culture reference (to a similarly posed centerfold of movie star Burt Reynolds in *Cosmopolitan*) may have been lost on Southeast Asian readers. "Thieu assumed that Kissinger had posed for the picture and was flabbergasted," wrote Walter Isaacson.[5]

The Presidential Palace made Nixon's emissary wait outside for a quarter of an hour, giving news photographers plenty of time to take real pictures of Kissinger. Inside it was worse. Kissinger faced not only Thieu, but his entire national security council, plus all of his ambassadors to the Paris talks. A room full of men who would move from high-ranking government positions into reeducation camps, prisons, graveyards, or exile under the "decent interval" terms Kissinger was there to extol. The South Vietnamese president offered no greeting. Thieu said that his press aide/nephew, Hoang Duc Nha, would translate. Kissinger looked at the Vietnamese men in the room; they all knew English.[6] (He despised "the egregious Nha," who was "educated in the United States and in the process had seen too many movies of sharp young men succeeding by their wits"—an odd criticism by a refugee who was educated in the United States and had succeeded by his wits.)[7]

Kissinger asked if he should speak. Thieu nodded without a word. For the next two hours, the national security adviser told a heroic tale of an embattled American president who had braved mounting political opposition at home and finally forced "a major collapse of the Communist position."[8]

Thieu listened intently, dragging on a long, thin Schimmelpenninck cigar.[9] He asked for a copy of the proposals in Vietnamese. Kissinger didn't have one. The South Vietnamese president asked cool, pointed, skeptical, but not hostile questions. If he decided that South Vietnam could not accept the agreement, did Nixon need to make it public before Election Day and announce his intention to sign it?

"I can answer you best by reading to you in part what the president delivered to me on the plane when I was leaving," said Kissinger, pulling out the note. When he read the line, "First, do what is right without regard to the election," Nha asked him to repeat that several times, expressing great difficulty understanding the words.[10]

Back at the American ambassador's residence, Kissinger received an unfortunate bit of good news: Hanoi had conceded the last two points. It agreed to let the United States replace South Vietnamese arms on a one-to-one basis. Nixon immediately launched Operation Enhance Plus, flooding Saigon with military supplies, so postsettlement replacement of used war material could proceed on a virtually unlimited basis.[11] Hanoi also agreed to let 30,000 civilian Vietcong supporters remain in

Saigon's prisons for the time being.¹² "I recognize this message adds immeasurably to your burdens," Haig wrote Kissinger.¹³ The deal was well-nigh done. Both Hanoi and Saigon now had motive to blow up—the former to make Nixon seal the deal, the latter to make him abandon it. Haig would try to buy some time by insisting that, in addition to the agreement, Hanoi make some "essential unilateral statements" on Laos and Cambodia.¹⁴

Kissinger cabled the White House that he couldn't "yet judge whether Thieu will go along with us."¹⁵

The next day, Thieu at least greeted Kissinger. "Good afternoon," he said in English.

The South Vietnamese president reviewed the proposed settlement point by point. It "only mentioned three countries of Indochina," said Thieu. That was the correct number only if South Vietnam was not really a country. Upon this point a war was being fought. "Our question," said Thieu, "is whether you consider there are three or four countries in Indochina."

Another extremely awkward moment. "That is a mistake," said Kissinger. He would have the number removed from the final text.

Kissinger tried to assure him on the key point: the continued presence of North Vietnamese troops militarily occupying and governing South Vietnamese territory. He tried to make Thieu believe what he and Nixon did not—that the security guarantees would lead to the closing of the Ho Chi Minh Trail. "The North Vietnamese forces in Laos and the Khmer Republic [Cambodia] must be withdrawn into North Vietnam," said Kissinger. If that were to actually happen, then no more soldiers and supplies would move down the trail into the South. "It is the judgment of our military people," said Kissinger, "that the present North Vietnamese forces without reinforcement will eventually have to be withdrawn or gradually whittled down."

Equally important as the settlement's provisions, Kissinger argued, was America's willingness to enforce them militarily: "President Nixon will not destroy in the last four years [of his presidency] what he has suffered for during his first four years. That is your guarantee of military survival. It is inconceivable that President Nixon would stand by if North Vietnam attacked again."¹⁶ Inconceivable. Left unsaid was the key change in political circumstance: In the first four years of his presidency, Nixon's reelection depended on Saigon's survival. Not so in the last four. The South had just about outlived its political usefulness.

Nixon had sent someone else to Saigon to give his assurances credibility: Gen. Creighton W. Abrams. The day Abrams was sworn in as

Army chief of staff, Nixon summoned him to the Oval Office and bluntly asked, "The question is, do you think you can help sell it to Thieu? In good conscience?"

"Oh, yes. I don't have any trouble with my conscience."

"You think it's the right thing to do," said Nixon.

"I guess," said Abrams. He spoke of progress. "In the beginning, they really didn't have any military forces. They didn't have the tradition of it, they didn't have the training of it, they didn't have the development of Vietnamese leadership. They lacked a whole lot of other things."[17] Abrams had helped build the South Vietnamese army to 1.1 million soldiers.[18] (American intelligence estimated that there were 150,000 North Vietnamese soldiers in the South; Saigon estimated 300,000. Either way, the South's army was several times the size of the North's.)[19] "Well, the time's gone on, and they've developed really quite a substantial military force with a lot of substance to it," Abrams said. "They've got a government. It doesn't always work in the ways that all Americans would agree with, but it is a government and it's done a lot for the country. A lot of it, of course, with our help, especially money. And so I think that this is another step in the process." It was not the most ringing endorsement. "You know, it's not all bad, us leaving Vietnam," said the general. "And I have to admit, too, that I—as far as I am personally concerned, [*unclear*] my thought has been for three or four years that it should never end with our military there."

Neither side would abide by the agreement, the president said. "The main thing is in fact who has power in place," he said. "And if Thieu's government and his military establishment is sound and strong, he'll survive. And if it doesn't, all the supervisory agents in the world are not going to assure his survival."[20]

In Saigon with Kissinger on October 20, Abrams told Thieu, "I [am] confident that the structure here as it stands today is capable of securing this country and this government." The structure at that moment still included more than 30,000 American soldiers on the ground. The settlement would remove them all. "I have always had great respect and admiration for the South Vietnamese people and military," Abrams said, "but I have always believed from the beginning that the day had to come for you and for your own pride and your people when the security and the political strength was all yours, with eventually our air power standing in the wings and our equipment and supplies coming into your ports." The general looked at Kissinger and said, "I think that's it."[21]

Kissinger's report to Nixon continued their joint infantilization of the South Vietnamese: "It was clear from the sober, somewhat sad, mood of the session that they are having great psychological difficulty with cutting the American umbilical cord." If the president had compared Saigon to a "child sucking at the tit," his adviser portrayed it as barely emerged from the womb. "Against this mood I did my best to underline their inherent advantages, draw out their self-confidence and assure them of US backing, both during an agreement and in the face of violations. I was partly, not totally, successful," wrote Kissinger. "I have the sense that they are slowly coming along and are working themselves into the mental frame of accepting the plan, but their self-respect requires a sense of participation."

While patronizing Saigon, Kissinger also managed to insinuate that the problem was, in part, one of American will: "They know what they have to do and it is very painful. They are probably even right. If we could last two more years they would have it made."[22] This implied that Nixon was leaving before the job was done (and that somehow two more years would do it when the previous eight had not). It was a subtle, hawkish attack, whether intended as such or not.

The president responded in kind:

> As you continue discussions with Thieu, I wish to re-emphasize again that nothing that is done should be influenced by the U.S. election deadline. I have concluded that a settlement which takes place before the election which is, at best, a washout has a high risk of severely damaging the U.S. domestic scene, if the settlement were to open us up to the charge that we made a poorer settlement now than what we might have achieved had we waited until after the election. The essential requirement is that Thieu's acceptance must be wholehearted so that the charge cannot be made that we have forced him into a settlement which was not in the interest of preventing a Communist takeover of a substantial part of the territory of South Vietnam.
>
> As I outlined yesterday, we must have Thieu as a willing partner in making any agreement. It cannot be a shotgun marriage. I am aware of the risk that Hanoi might go public but am confident that we can handle this eventuality much easier than we could handle a pre-election blow-up with Thieu or an agreement which would be criticized as a pretext for U.S. withdrawal.[23]

Kissinger would take offense at this in one of his memoirs ("Nixon ironically was implying the same charge against me that others were to make against Nixon—that I was rushing it for the election") and portray himself once more as above politics ("Of all of Nixon's assistants

I was the least involved in the election campaign"). By necessity, the memoirist ignored the real teeth Nixon had bared in the cable: Kissinger *was* negotiating "a pretext for withdrawal" that would lead not only to "a Communist takeover of a substantial part of the territory of South Vietnam" but to the loss of all of it. Nixon was reminding Kissinger—as Kissinger had recently reminded the president—that Thieu was right. Kissinger couldn't acknowledge in his memoirs the substance of Nixon's subtle, hawkish counterattack without giving away the game. He did write that, after reading the above cable, "I began to be nagged by the unworthy notion that I was being set up as the fall guy in case anything went wrong." Replace "in case" with "when" to appreciate the depth of his fear. The success of a "decent interval" exit strategy requires a fall guy.[24]

"No Possibility Whatever"

Foreign Minister Tran Van Lam invited Kissinger into his Saigon home on October 21 and, during an opening prayer, expressed thanks for his visit and work before asking the Lord to bless the Vietnamese and American people at this crucial moment.

Thieu was not present, but his foreign affairs adviser, Nguyen Phu Duc, politely went through the draft and offered "suggested changes." Twenty-three of them. When he got to the part about total American withdrawal, he said, "We suggest adding another paragraph: 'Concurrently, North Vietnamese regular and irregular forces will be withdrawn from South Vietnam into North Vietnam.'" A far from outlandish suggestion, President Nixon himself having publicly demanded no less during his first three years in office.

Kissinger was blunt: "I believe there is no possibility whatever of getting them to agree to this."[1]

Thieu had a meeting scheduled with Kissinger that afternoon. He postponed it several times, then canceled altogether. Nha ("the egregious") called to say the South Vietnamese president would see him the next morning at 8:00 a.m.[2] Meanwhile, Hanoi crushed all hope for delay by mercilessly meeting Haig's demands on Laos and Cambodia.[3]

"It now appears that your meeting with Thieu is a decisive one," Haig cabled. Nixon, unhelpfully, directed Kissinger both to push Thieu to the limit and to avoid "forcing him to break publicly with us before November 7." More helpfully, the president wrote a threatening letter for him to brandish at the 8:00 a.m. meeting.

"The Man Who Should Cry Is I"

Thieu stalled. He told Kissinger at the morning meeting that he would give him a final answer that afternoon. South Vietnam could not, however, accept two things: (1) the continued presence of North Vietnamese troops and (2) the equal treatment of Communists, anti-Communists, and neutrals on the election commission. These would be opposed by all South Vietnamese except for a small minority, said Thieu—adding that the minority would be eliminated.

Kissinger tried umbrage: never had he, a representative of the president, been subjected to such treatment as he had in Saigon.

Kissinger tried statistics: Thieu's army outnumbered the North's by 11 to 2. The numbers favored the South. They often had over the last decade. But the numbers did not measure the will to fight.

Kissinger tried a threat, personally penned by the president: "Were you to find the agreement to be unacceptable at this point and the other side were to reveal the extraordinary limits to which it has gone in meeting demands put upon them, it is my judgment that your decision would have the most serious effects upon my ability to continue to provide support for you and for the Government of South Vietnam."

Kissinger tried a different threat: if the draft agreement became public, Congress would certainly cut off aid.

Kissinger tried to unite with Thieu against a common enemy: the American news media. The press, said Kissinger, had a vested interest in defeat in Vietnam, being violently opposed to both President Nixon and President Thieu. He tried an argument more suited to the former than the latter: an agreement would confound the press and the liberals!

Thieu remained polite, controlled. He would give his answer at 5:00 p.m. He said he was trying to avoid a situation like 1968. If this was a threat, it was veiled.[1]

"I think we finally made a breakthrough," Kissinger cabled the White House.[2] Between meetings he flew to Cambodia to brief Thieu's counterpart, President Lon Nol.

Response to the proposed settlement in Laos and Cambodia had been extremely favorable, Kissinger told Thieu on his return. (Word that the North would finally close the Ho Chi Minh Trail and withdraw from the two countries was welcome, if untrue.)

They are not being sacrificed by their allies, Thieu replied. The South Vietnamese president was ready to show his hand. Thieu called the election commission a disguised coalition government that would lead to

South Vietnam's collapse. Clearly, the United States, the Soviets, and the Chinese had decided there were only three countries in Indochina, and South Vietnam wasn't one of them. "I have a right to expect that the U.S. has connived with the Soviets and China," said Thieu. "Now that you recognize the presence of North Vietnamese here, the South Vietnamese people will assume that we have been sold out by the U.S. and that North Vietnam has won the war." The English translation was slightly off, but the geopolitical and political analysis was spot-on. Kissinger told him that Le Duc Tho had burst into tears, and Thieu replied, "the man who should cry is I." And he did, more than once during the meeting.

Kissinger took as much umbrage at the accusation as he could muster: "As an American, I can only deeply resent your suggestion that we have connived with the Soviets and the Chinese." (As someone who had connived extensively with both Communist powers to obtain a "decent interval," on the other hand, Kissinger could have chosen to respond with something other than resentment.) "How can you conceive this possible when the President on May 8 risked his whole political future to come to your assistance?" (The poll-tested bombing and mining Nixon announced in May was the opposite of a political risk.) "When we talked with the Soviets and Chinese, it was to pressure them to exert pressure on Hanoi." (Yes, to pressure Hanoi to delay the overthrow of Thieu's government for a "decent interval.")

Kissinger termed Thieu's position suicidal.

"If we accept the document as it stands, we will commit suicide," Thieu said (correctly, as Kissinger had advised his president earlier that month in the Oval Office).

The South Vietnamese president ended the meeting by letting Kissinger know that he would see him once more before Kissinger went home to Washington.[3]

Kissinger sent a cable to greet Nixon when he awoke: "Thieu has just rejected the entire plan or any modification of it and refuses to discuss any further negotiations on the basis of it." Thieu wanted "total withdrawal of North Vietnamese forces, and total self-determination of South Vietnam without any reference as to how this is to be exercised. I need not tell you the crisis with which this confronts us."[4] (Nixon had already asked Haldeman for a poll on the issue of a ceasefire-in-place versus North Vietnamese withdrawal from the South.)[5]

While Nixon slept, Kissinger tried to delay the detonation of two political time bombs. He dispatched a message to Hanoi in Nixon's name, blaming the North for a "breach of confidence." North Vietnamese premier Pham Van Dong had given an interview on the negotiations to

Newsweek. He spoke of a "three-sided coalition of transition" for the South. This made the election commission sound like a coalition government. It would help Thieu make a case against the settlement in American public opinion, Nixon having blasted the coalition concept for years as surrender on the installment plan. The *Newsweek* interview, Kissinger wrote as Nixon, "bears considerable responsibility for the state of affairs in Saigon."[6]

As he bade farewell, on three hours' sleep, to the South Vietnamese president in the morning, Kissinger said he had cancelled his trip to North Vietnam. He portrayed it as an opportunity lost: if he had gone to Hanoi, the Communists would have made more concessions, and the agreement would have been viewed as a triumph. "Had it not been for the importance we place on our relationship, we would not have to make new plans," said Kissinger. "This is why I leave with such a sense of tragedy." He emphasized unity, expressed as secrecy: "We must let no disagreement be evident between us."

Thieu spoke in the past tense. "I promised to avoid any confrontation and said that I would not publicly acknowledge any disagreement between President Nixon and myself," he said. "I still consider President Nixon a friend and a comrade in arms." He didn't say what he thought of Kissinger. (His scribbled notes from the previous day included the words "errand boy" and a smattering of curses in English and French.)[7] Thieu defanged any threat to oust him in a coup by saying that "whether or not I am president I will strive to create conditions so that the United States can help Vietnam. If I am an obstacle to American aid or to peace, I will not stay on as president."

"The U.S. will never," Kissinger said, "sacrifice a trusted friend."[8] (He was less politic with Commerce Secretary Pete Peterson back in Washington: "I have only one desire—to turn the Vietnamese loose on each other in the hope that the maximum will kill each other off.")[9]

"The Fellow Is Off His Head"

October 23, 1972, 11:22 p.m., National Security Adviser's Office, the White House. Kissinger was keeping the president waiting. Before he could join Nixon, Kissinger first had to connive. "I just got in," he said into the phone.

"Yes," said Soviet ambassador Anatoly Dobrynin, "so I imagine."

"Time is of some importance," said Kissinger. "We received a somewhat threatening message from Hanoi."

"Yes."

"They don't understand the American domestic situation at all," said Kissinger. "What we cannot do is to have a confrontation with the South Vietnamese in the last two weeks of the elections and be accused of undermining the people for whom we've been fighting for four years. Now that is a fact of life."

"Yes." Dobrynin understood. Before Kissinger had left for Paris and Saigon, he had enlisted the Soviet ambassador's aid in persuading Hanoi to withdraw some of its ten divisions from the South. "You mean," Dobrynin said, "very much symbolic to begin with."

"Yes," Kissinger said then. Now he renewed the request with urgency.

"The major issue is to do something about their troops. Not all of them," Kissinger said. "And it's not to be written in the agreement."

"Yeah, I understand."

"They may think we are deliberately delaying it beyond November 7th so we can bomb them or something. I give you the solemn assurance of the president that this is not the case."[1]

Down the hall in the Oval Office, waiting for Kissinger to get off the phone, Nixon told Gen. Alexander Haig the opposite: "After the election, we'll bomb the bejeezus out of them. I'm not going to tell Henry that, but that's how we're going to do it. He's got to go another round with them."[2] (Nixon had been unusually frank with Haig in Kissinger's absence. That morning, the president had told the general: "Call it cosmetics or whatever you want to do, this has got to be done in a way where we give South Vietnam a chance to survive—and not to survive forever, but they've got to survive a reasonable time. And if they don't, everybody will say, well, goddamn it, we did our part.")[3]

"Henry feeling better?" the president asked Haig.

"Yes," said Haig. Kissinger was tired but feeling very good. "Henry had a hell of a time with Thieu. He's a paranoid of the first order," said Haig. "We're just going to have to get the hell out of there. We've done all that has to be done. If he can't cut it with a million men with all that equipment then goddamn it, he wasn't worth saving in the first place."

"That's right," said the president. "He can't cut it."

"He'll cut it all right, because they're on their knees," said the general. Hanoi wanted to settle. "I'm personally relieved that we don't have to do this before the election, because it's going to be messy," said Haig. "It couldn't help. It could only hurt."

"Right now, we don't want to hurt," said Nixon. "We just want to keep the thing confused and fuzzed up."

"And that's easy," said Haig. Kissinger could brief reporters.

"Got to use the word 'progress,'" said Nixon.

"Right."

"So that we can give hope," said the president. Not that he needed it. "The election is done," said Nixon. "And it's a good thing. That son of a bitch has got to be crushed."

"That's exactly right, Mr. President." Voters preferred Nixon to McGovern on Vietnam.[4] "This actually has turned out very well. It's been good, solid, steady progress in search of peace. No one doubts that you have made every effort," said the general.

"That's right. That's for sure." Of course, Hanoi might go public with the draft settlement. The president already knew how he would handle it if Hanoi revealed the truth. "I will say that they have reneged," said Nixon. "Are the American people going to believe them or us?" Afterward, "we're going to bomb the shit out of them. I really mean we'll bomb them this time. No problem."[5]

The national security adviser finally joined them. He accentuated the positive: "What was, I think, a good agreement then, has become immeasurably better."

"Oh," the president said, "I know."

"They've withdrawn their demand that Thieu had to promise to resign some time before the election. In other words, if we compare the agreement with what you said you would do on May 8th, we have almost exceeded it," said Kissinger.

"We have more guarantees," said Nixon.

"And these guys, after having stonewalled us for 10 years, have caved on every single issue," Kissinger said. "It's the very best we could have gotten." The problem was Thieu. "And I'm afraid that the fellow is off his head," said Kissinger.

"He wants us to be in there," said the president. "Well, the problem that I see with the whole thing is, looking at it coldly on the merits, who else?"

"If we overthrow him now, first of all, he's so insane that he'll take everything with him," said Kissinger.

"He's cracking under the strains," said Nixon.

"The only rational explanation I have for what he did—which he didn't make, but which I would have to make on his behalf—is, if he has three weeks, he can get his army deployed, he can get his province chiefs lined up, and he can have a good—a better chance to survive the ceasefire. Secondly, if he doesn't yield to me when I come there, then he's not an American stooge. He's done it on his own. And if he does it in a shitty way, we can swallow our pride," said Kissinger.

Another rational explanation for Thieu's behavior—one provided to Kissinger that day by John Negroponte, his Vietnam expert—went unmentioned. The fundamental issue, Negroponte wrote, was what the Communist forces in the South would do when the ceasefire left Saigon's forces "frozen in place." Negroponte was certain that Hanoi would reinforce Communist forces "with an imperceptible trickle of men and supplies down the infiltration corridors" of the Ho Chi Minh Trail through Laos and Cambodia. "Saigon's nightmare," Negroponte wrote, "is not the absence of unenforceable legalistic clauses about [the North Vietnamese army's] withdrawal, but rather that this treaty would recognize the presence of 15 NVA division equivalents in the South." The Communists' presence on the election commission, moreover, gave them political legitimacy. "Politically, they will be permitted to operate legally in Saigon; militarily, they are certain to operate illegally in the boondocks." The commission was sure to deadlock, and that "will justify renewed military activity by the other side, if it hasn't already occurred by that time." Most South Vietnamese leaders would view "precipitate US disengagement," coupled with the agreement's military and political restraints, as "the handwriting on the wall." Negroponte's memo was timely, prescient, and classified for more than a quarter of a century.[6]

"But I don't exclude the irrational explanation," Kissinger continued, "that after the election he'll be just as miserable and paranoid as he is."

"But then so will we," said the president.

"Mr. President, I'm glad that we're—that we now have to do it after the election, unless you absolutely had to have it before the election—"

"We don't." He'd been campaigning that day in New York with the state's Republican governor, Nelson Rockefeller, Kissinger's previous patron—"waving my arms for four hours in a goddamn motorcade to half a million people" in the suburbs. When they were alone, Rocky had advised him against making a deal now. "'Look,' he said, 'it's too close to the election. It'll look like an election trick.'" Nixon's old rival had some great news: a poll showed him winning the state by 58–32 percent. "This asshole McGovern—I hope I don't bother you with it—but he has been the most irresponsible son of a bitch," said Nixon. "On the one hand, he says we're going to dump Thieu in order to have an election settlement. On the other hand, he says, Thieu's blocking it." The president's outrage was undiminished by the facts that (1) the chief reason he wasn't dumping Thieu was his need for Saigon to last a "decent interval," and (2) Thieu was indeed blocking the settlement. Nixon asked about Ambassador Dobrynin.

"I gave him this message: I said, 'If the President has to choose between a blow-up with Hanoi and a blow-up with Saigon this time, he'll unhesitatingly choose a blow-up with Hanoi,'" Kissinger reported. "Therefore, don't threaten us."

"That's right."

Moscow could explain to Hanoi "that we cannot take on Saigon before November 7th. But if they go through one more negotiating round with us, and if they give us the opportunity to present the concerns of our allies, and if we can then get some of these changes made, we will consider it a final document," Kissinger said. Saigon "even gave us 23 changes. Of those, 16 are just crap."

The president returned to his primary concern. "Right now, Henry, we've got to crush McGovern. He's a bad man. He's got to be crushed bad. Now, that's point one," said Nixon. "Second, if you could keep the thing confused for a while." Kissinger might have to hold a briefing. "You go in and you know how to bamboozle the bastards anyway," said Nixon. The president tossed him some lines: "'We have no election deadline.'"

"That's right."

"'The deadline only is to do the right thing at the right time,'" said the president.

"I can say we've made major progress."

"Oh," said Nixon, "use that word." As they discussed the words Kissinger would use to defend the agreement, Nixon's mind turned back to the obstacle: "What's he want—he wants us in forever?"

"I have come to the reluctant conclusion, Mr. President, and it breaks my heart to say it, that that system in South Vietnam is geared to a war which we sustain and that these guys cannot imagine what peace would be like. And that they're terrified not so much of the Communists. They're terrified of—of peace," said Kissinger.

"Incidentally, did Abrams behave?" asked Nixon.

"He was very good. The first day, he didn't say anything. The second day he made a very good speech," said Kissinger.

"He put himself solidly behind the agreement?"

"One hundred percent behind the agreement." The leaders of Laos and Cambodia were for it, Kissinger said, unlike Thieu: "That son of a bitch says you have been plotting with China and Russia to do him in for a year."

"China and Russia, for Christ's sakes, have been our great allies at this point. They haven't been screwing, they've been helping us," said Nixon. (That was what Thieu was afraid of.)

"Since I think now that these hysterical bastards in Saigon may collapse on us, it is better that you've done it after the election than that you've done it as an expedient before," said Kissinger.

"God, if they collapse, what happens then?"

"Well—"

"Everything has been in vain," said the president.

"I think our honor is—is intact. They won't collapse that fast," said Kissinger.

"That's my view. I don't believe they'll go that fast," said Nixon. They talked a while more before the president looked at the time. "That clock up there is wrong, isn't it? It's not 9:30. It's about 1:30, isn't it?"

"It's exactly midnight," said Kissinger.

"Well, it's a very good time for you to go to bed," said Nixon.[7] He'd had a long day of campaigning.

It had also been Veterans Day. The president did not attend ceremonies honoring veterans at Arlington National Cemetery, but he did send a prerecorded message to play there among the patriot graves: "As American troops return home with honor from another distant conflict, the hope is strong for a full generation of peace. No group has sacrificed more to achieve this goal than the men and women who have proudly worn the American uniform. . . . To all of them I say that our respect has never been stronger, nor our gratitude greater, than on this Veterans Day 1972."[8]

Before going to bed, Kissinger launched himself into a vigorous round of triangular diplomacy with a focused goal: keeping the North Vietnamese from revealing the settlement he had reached with them until after Election Day. This was a challenge. For one thing, Nixon was unwilling to stop bombing them, even though they had met his demands. He had the issue polled: public opinion in the fall of 1972 ran against a bombing halt, 82–18 percent.[9]

"It is a hell of a thing to ask Hanoi to—that it must be bombed because Saigon doesn't accept their terms," Kissinger told the president.

"Keep the bombing thing in there as a card to give up," Nixon replied.[10] In return for what, he did not say. He did, however, agree to bomb *less*.

"I've just talked to the president," Kissinger told Ambassador Dobrynin at 12:10 a.m. "We, as a sign of our good faith, are going to stop bombing north of the 20th parallel."

"OK. I think it is, too, rather important," said the Russian. "I'm sending right now a telegram."[11]

Ambassador Huang Hua, representative of the People's Republic of China to the United Nations, agreed to see Kissinger that day in New

York. "You must be tired," said Huang. "You spend half the time on the ground and half the time in the air."

"After my last stop, I enjoy being in the air more than being on the ground."

The ambassador smiled slightly.

Kissinger tried again: "I have achieved the unity of the Vietnamese—both of them dislike me, North and South." The ambassador laughed. "I don't want to take too much of your time," Kissinger said, "but I asked to see you shortly after my return because I want to ask something which we have not asked before—and that is whether the Prime Minister might be willing to use his good offices in the rather complicated state that our negotiations have reached with the Vietnamese." While he was in Saigon, it had "not proven possible to obtain agreement in every respect," and "some aspects of the agreement" had to be "slightly adjusted without major changes." Nevertheless, Hanoi insisted that it be signed by October 31. "But this is insanity," Kissinger said. "Mr. Ambassador, you know the United States. You know that between now and elections we cannot have a public confrontation with Saigon." He requested Zhou's assistance "to convince Hanoi that this is not a trick." Many of the changes Saigon wanted would be quickly forgotten once the agreement was signed, said Kissinger. "It is merely a question of face." Ambassador Huang's note taker and his interpreter appeared not to understand the last word. The ambassador repeated it for them. "Someone once told me," Kissinger said, "that westerners are conscious about face."

As if to demonstrate this point, the national security adviser proposed that in response to Saigon's demand that the North Vietnamese remove all their troops from its territory, "they should withdraw some forces from the northernmost part of the country, 20 kilometers to North Vietnam." This wouldn't "change the military situation very much, because it is very close, but it would satisfy the political requirements of the situation." In other words, a token, face-saving withdrawal.

Kissinger also wanted a different Vietnamese translation for the election commission, to make it absolutely clear that it wasn't a government.

"The issues are not major ones," said Kissinger. "Now if Hanoi makes a public issue of it, we will be forced to emphasize all our differences and a settlement would be delayed indefinitely."

Ambassador Huang dropped his matches.

"We thought that if someone could make clear that our tendency is to keep our promises," Kissinger said, "this would have a helpful influence."

"If there is nothing else you would like to tell us, we will take our leave," said the ambassador. As the Chinese awaited their car, Ambassador Huang asked if there was any way to complete the settlement before the election. Kissinger's aide summarized his response: "In a blow-up before the elections, the U.S. would have to choose Saigon over Hanoi. After [the] elections, it would be just the opposite."[12]

Almost two hundred Vietnam veterans watched the president that day as he formally signed a bill that increased their educational benefit. "Some of you have paid a very high price, I see several in wheelchairs here," he said before announcing that everyone present would get a souvenir pen. "I trust that you can endorse the first check with that increased benefit with it!" After the ceremony, a veteran on crutches worked his way over to where Nixon stood. He shook the president's hand. Larry Kirk. Army. Lost both legs below the knees in Vietnam, and part of one arm. He hoped the president meant what he said about giving veterans "the first crack" at jobs, since he'd applied for one as a White House Fellow. Very competitive program—an opportunity for the best and the brightest to work for a year in the White House or a Cabinet department. The Jaycees had honored Kirk on its annual list of Ten Outstanding Young Men.

"You want to be a White House Fellow?" the president asked. "Seriously?"

Kirk said he wanted to "continue to serve my country."

"It's arranged," said the president. A press photographer captured the moment: Kirk looks relaxed and friendly, but modest; Nixon . . . angry. This spontaneous, heartwarming anecdote made the *New York Times, Washington Post,* and a lot of other papers, thanks to an Associated Press dispatch.[13] Seven months later, the White House announced the selection of eighteen Fellows for the coming year; Kirk's name wasn't on the list.[14]

The president told his diary:

> As I spoke to the veterans I was again terribly moved when I saw the blind veterans and those in wheelchairs.
> It makes me realize what a debt we owe to these people, and how important it is to have the kind of peace that will really contribute to no more wars, rather than one that is simply a hiatus between two wars.[15]

No Coalition Government

CBS, NBC, and ABC all had the same top story on the evening news: a hard-line television speech by South Vietnam's president setting his conditions for a settlement: no North Vietnamese troops in the South, no coalition government. "Thieu didn't say so in his television broadcast," NBC's Tom Streithorst reported from Saigon, "but during his five-day visit here, presidential adviser Henry Kissinger reportedly pressured South Vietnam to accept some form of coalition." Saigon was calling the election commission a coalition government—a concept Nixon himself had assailed as a "thinly disguised surrender."[1] Thieu was staying within the letter of his commitment not to make his differences with Nixon public. Publicly, he spoke in general terms. All the peace proposals discussed in Paris, he said, were unacceptable.[2]

McGovern pounced. A settlement now would mean Nixon had added four years to the war "purely to avoid criticism from the right-wing war hawks," said the Democrat. This was the closest he had ever come to identifying Nixon's actual strategy.

Standing on the steps of a county courthouse in Dayton, Ohio, hatless in a sudden shower, McGovern asked, "Why, Mr. Nixon, was it necessary to kill another 20,000 young Americans in this war before we end it?

"What did you gain by killing or wounding or driving out of their homes six million people, most of them in South Vietnam, by this incredible bombing that has gone on for the last four years?

"What did you get, Mr. Nixon, for the $60 billion you spent in the last four years on the destruction of Southeast Asia that we needed to build up our own cities, to combat pollution, to build up our own country instead of destroying the land and the villages of another country 10,000 miles from our shores?

"I ask this question: What has changed that makes it any easier for us to get a peace settlement today than the one we could have had four years ago, if we had a president committed to peace four years ago?

"Did you make all these sacrifices, Mr. Nixon, to save your own political face from right-wing criticism?"

McGovern's attack was based on the premise that Nixon had agreed to a coalition government. If true, "any settlement that comes now in the closing days of this election campaign is the same kind of settlement we could have had four years ago," McGovern told *CBS Morning News* in New York. Such a settlement "would destroy Mr. Nixon."[3]

If a coalition was a "disguised surrender," and Nixon were agreeing to one now, then he had prolonged the war for his entire first term only to abandon the South when politically convenient. McGovern didn't raise the possibility that the president might be settling for a different kind of disguised defeat.

Rumblings about a coalition government forced Nixon to adjust his campaign rhetoric. In eight separate speeches during the campaign's final days, the president made the same claim in almost the same words: the North Vietnamese "have agreed that the people of South Vietnam shall have the right to determine their own future without having a Communist government or a coalition government imposed upon them against their will."[4] The coalition half of this line was true.

Woodward and Bernstein's biggest Watergate story to date appeared on the front page of the October 25 *Washington Post:* "H. R. Haldeman, President Nixon's White House chief of staff, was one of five high-ranking presidential associates authorized to approve payments from a secret Nixon campaign cash fund, according to federal investigators and accounts of sworn testimony before the Watergate grand jury."[5] Haldeman's authority in the White House was second only to the president's. That he had a hand in controlling the secret fund for the spying and espionage campaign against Democratic presidential candidates brought the Watergate story into Nixon's inner circle.

White House Press Secretary Ron Ziegler devoted a half hour of his daily news briefing to denouncing the *Post*. This time he had backup. The attorney for Hugh Sloan, former treasurer of CREEP, denied that his client had named Haldeman before the grand jury. (Sloan did name Haldeman under questioning by Woodward and Bernstein; the grand jury hadn't asked.)[6] The full-throttle denunciation by Ziegler and the technical basis for the White House denial showed how seriously the president took the *Post*'s investigation in the closing days of the campaign.

All three networks ran the Haldeman story and the denial. But the next day the evening news would have no time for Watergate.[7]

"Peace Is at Hand"

"We've had another very major development," Kissinger reported to the president on October 26, 1972, "which we haven't had a chance to brief you on."

"Oh. OK. Fine."

"Uh," Kissinger said. And. Paused. For. Five. Seconds. "Uh, [North Vietnam] has gone public." Radio Hanoi was broadcasting the terms of the settlement. The Communists claimed that "so-called" difficulties with Saigon were just an American excuse to postpone peace. "And they demand that we sign it by October 31st as we have promised."

Nixon refrained from blowing up. He already knew; Ehrlichman had told him. Not that Nixon mentioned that. He let Kissinger squirm a little. The president asked, "Well, what do we have to do?"

"Well, what I thought I'd have to do is—"

"Brief?"

"—brief this morning."

They had decided this in advance, but now, with Kissinger about to enter the spotlight on the biggest issue of the campaign, the president acted hesitant. "Well, now wait a minute. Let's just think what we're doing. What's the purpose of the briefing?" asked Nixon. "That we set it straight?"

Kissinger auditioned. He rehearsed what he would say to the press. Nixon fed him lines to use: "We shall never agree to imposing a Communist government on the people of South Vietnam." "The important thing is not whether it comes before or after the election, but that it be the right agreement." "We are not going to have the sacrifices of the American people to prevent the imposition of a Communist government in South Vietnam [*unclear*] to have been in vain."

"I'd just give them a little of that purple rhetoric if you could," the president said. If they asked about the bombing, Nixon suggested saying, "'You will note, gentlemen, we are continuing the mining.' I'd put that in."

"What I wanted to ask you, Mr. President, is, should I say that we offered to have one more meeting with them to clean up the text, and that then we would reduce the bombing?"

"No. No, no, no." Reports that Nixon had scaled back the bombing to the twentieth parallel were "harmful," the president said. "Don't say it publicly. It'll kill us, utterly kill us."

"All right," said Kissinger.

"It's very, very bad," said Nixon. "You could be conciliatory toward Hanoi, but I would treat McGovern and his people with utter contempt."

"Oh, yes," said Kissinger. (He would neither attack the Democrat nor so much as mention his existence. This was Kissinger's moment in the spotlight; he would appear to be above politics.)

"Incidentally, you going to let television cover you?" Nixon asked, as if it were up to Kissinger. For four years the White House had kept his voice off the airwaves as much as possible. The stated reason was that no one knew how a German accent would go down with Middle Americans who remembered World War II (or World War I).

"What do you think?" asked Kissinger.

"I would," said Nixon. "It doesn't bother you, does it, the television?"
"No."[1]

By 11:30 a.m. every seat in the White House Press Room was filled. Overflow reporters crowded the aisles, lined the walls, hunched over the platform at the foot of the podium. "Ladies and gentlemen," Kissinger said. "It is obvious that a war that has been raging for 10 years is drawing to a conclusion." He was clearly new at this. His tie hung off-center. He licked his lips. Even worse, he kept clearing his throat every few seconds, with a sound like he was stretching out the last two letters of Bach's name: *ccchhhh*. "The President thought that it might be helpful if I came out here and spoke to you about what we have been doing, where we stand, and to put the various—*ccchhhhh*—allegations and charges into perspective. *Ccchhhh*. We believe that peace is at hand. *Ccchhhh*."

He touched every base. He mentioned that Hanoi had dropped its demand for a coalition government. He explained away South Vietnam's resistance to signing, saying, "It is inevitable that in a war of such complexity that there should be occasional difficulties in reaching a final solution." He ran through the provisions, highlighting the popular ones (release of the POWs, ceasefire, American withdrawal), glossing over the less popular ones (North Vietnamese military and political control over parts of the South came out as: "the existing authorities with respect to both internal and external politics would remain in office"), and managing to keep a straight face through the ones that were purely for show ("there is an affirmation of general principles guaranteeing the right of self-determination of the South Vietnamese people and that the South Vietnamese people should decide their political future through free and democratic elections. . . . [T]he [tripartite election commission] would operate on the basis of unanimity. . . . [T]hey agree to refrain from using the territory of Cambodia and the territory of Laos to encroach on the sovereignty and security of other countries. . . . [F]oreign countries shall withdraw their forces from Laos and Cambodia"). He spoke vaguely of America contributing to the reconstruction of Indochina but assiduously avoided the word that Nixon used in private, "reparations." He denied political considerations.

Reporters questioned the timing, so close to Election Day. What assurances could Kissinger give that the deal would not fall apart after the voting?

"*Ccchhhh*. We can only give the assurance of our record. We have conducted these negotiations for four years—*ccchhhhh*—and we have brought them to this point with considerable difficulties and with considerable anguish," Kissinger said. "And we cannot control if people believe or if people choose to assert that this is simply some trick."

Question: "Did President Thieu go along with this whole deal?"

"The South Vietnamese agreed with many parts of it and disagree with some aspects of it," Kissinger said. "And we agreed with some of their disagreement and not with all."

Question: "What are the main differences between Saigon and Washington now?"

"No useful purpose would be served—*ccchhhhh*—by going into the details of consultations that are still in process."

Question: "What concessions did the United States make to get this agreement?"

"The United States made the concessions that are described in the agreement," Kissinger said. He also answered a question that no one had asked: "There are no secret side agreements of any kind."[2]

The three networks literally raced to air Kissinger's words. CBS had the technological edge: an innovative, miniaturized camera that electronically transferred visual images onto magnetic tape and weighed only sixty pounds. Once a tape was full, a motorcyclist rushed it from the White House to the network studio. CBS managed to get Kissinger's press conference on the air a mere twenty-five minutes after he finished speaking, while ABC and NBC were still developing their film. All the networks ran specials on the announcement. ABC's was titled "Peace Is at Hand."[3] The stock market took wing. *Newsweek* would place a picture of a soldier on its cover, his helmet emblazoned "Goodbye Vietnam."[4]

Shortly before midnight, Kissinger's phone rang.

"Mr. President?"

"Well, Henry, I understand that all the three news shows were about Vietnam then. I wonder why."

"Heh-heh. Well, Colson called me and he thinks that we've wiped McGovern out now."

"Did he really?" Nixon had complained to Haldeman about Kissinger's performance: "The P felt that we were getting the wrong twist on this, and that K was getting the play, and that the announcement had been blown, where the P had hoped that he could go before the nation and make the announcement."[5]

Kissinger asked about the president's campaign rallies in West Virginia and Kentucky. "Good god, they practically [took] the roof off," recalled Nixon. "I suppose the problem we've got, which we have to bear in mind, is that it's just a week early." He recalled the last week of the previous campaign. "You remember in '68, after Johnson announced a bombing halt, Thieu blew it."

"Yeah."

"Blew the whistle."

"He hasn't blown the whistle this time," Kissinger said. "You got a good reaction."

Laughing, Nixon said, "Of course, the point is, they think you've got peace."

"Yeah."

"That's what they think, but that's all right. Let them think it."

"Essentially true, Mr. President."[6]

As soon as he hung up, Nixon called Chuck Colson.

"Well, you've had quite a full day, Mr. President."

"I was going to say we sort of knocked Watergate out tonight, didn't we?" asked the president.

The operative chuckled. "Well, the beauty of it was that the North Vietnamese put it out," said Colson. "We look tough and hard-line and the North Vietnamese look like they're coming to us." And "old George McGovern" looked like he had "a very bad case of indigestion."[7]

McGovern had been primed to attack if Nixon agreed to a coalition government. That would have given him the golden opportunity to say that the White House could have achieved as much four years earlier. Now that the world knew that Hanoi had dropped the coalition demand, McGovern faltered.

Rather than question the settlement terms, McGovern expressed "my hope and my prayer that these reports will turn out to be true."[8] He told reporters, "Whatever their motives, if the administration can bring off a settlement of this war, they'll have my full support and cooperation in any effort that can lead to peace."[9]

At the University of Iowa, the Democrat compared the settlement to the 1954 Geneva Accords. In that earlier agreement, Ho Chi Minh committed to withdraw his armed forces from the South and did so. He also agreed to submit to a nationwide election to choose a government for all of Vietnam. (American observers predicted that Ho would win that election as the leader of the victorious anti-French revolution; when Saigon's leaders refused to take part in such a vote, President Eisenhower sided with them.) With a ceasefire-in-place and elections that would occur only if Communists and anti-Communists agreed unanimously on how to run them, the terms Kissinger outlined were blatantly inferior to those of the Geneva Accords. When McGovern said that the terms Nixon had agreed to "are very similar to those that the French accepted back in 1954," he probably didn't mean it as extravagant praise.[10] What he did mean, however, was not clear.

McGovern repeated his rhetorical questions from before Kissinger's press conference—somewhat less effectively, now that the nation knew the answers the White House was giving: "What did either we or the rest of the world gain by the killing of another 20,000 young Americans these past four years?" For one thing, Hanoi's agreement to drop its demand for a coalition government. One could argue that the settlement would lead to resumed warfare between North and South, that the latter was much more likely to lose thanks to the requirement for complete withdrawal of American forces, and that Nixon's terms were, therefore, even worse than a coalition government. But McGovern did not make this case.[11]

Colson reviewed McGovern's performance for Nixon: "When he was asked about it, God, he looked—he looked pretty feeble."

With a few hours of perspective, Nixon concluded, "It's better that Henry announced this than I did."

"Yes."

"If I'd done it, it'd look like I was doing it as an election trick," said Nixon.

"Henry is not looked upon as a political fellow at all," said Colson. "I mean, he's looked upon as a professor, really. And more than that now. A great deal more than that now."[12]

Conservatives showered Kissinger with praise. The intellectual Right, in the form of William F. Buckley's brother James, a United States senator elected by New York in 1970 on the Conservative Party line, called to wish Kissinger "the Nobel Peace Prize."

"I don't know, but I think we're going to make it," said Kissinger.

"May I assume that what has been done will not pull any rugs out from [under] any regime?" Buckley asked.

"Absolutely, totally, 100 percent."

"Great," said Buckley.[13]

The moderate Right, in the form of UN ambassador George H. W. Bush, called to profess "adulation, I'm in your fan club."[14]

The populist Right, in the form of Gov. Ronald Reagan, called to quip, "Henry, Nancy and I were thinking of taking a few days off for a little rest and we've always wanted to see Hanoi. Would it be all right to leave, say, Sunday?"

Kissinger laughed.

"Listen, this was just wonderful," said Reagan.

Kissinger placed the blame for reports of a coalition government on "the liberal press."

"Well, then, there's nothing to all these stories that he is sitting there balking?" asked Reagan.

"Well, he is going through a complicated maneuver trying to prove to his people that he's not an American stooge," said Kissinger.

"Uh-huh, yeah." During the press conference, Reagan had been at a television station where "they had some prisoner of war wives who were watching, and I went in and spoke to them, and saw them, and I want to tell you, they were smiling through tears." Even a TV crew guy ("I'm quite sure a Democrat") said Kissinger was brilliant. Reagan seconded that assessment.

Kissinger thanked him: "It was the support of people like you in the dark periods that made it possible."[15]

For conservatives, it was a time for choosing. If they chose to believe Kissinger, then they could see themselves as strong, firm, steadfast Americans whose support for Nixon and the war had finally been vindicated in the face of all the naysayers. They could be heroes in their own eyes.

But what if the South Vietnamese were right? If conservatives listened to the people they longed to think they were saving, then they'd have to see themselves in terms they despised—as people who had supported a president who had waged a no-win war until a politically convenient time arrived to sell out an ally.

The Democrats were hardly more reserved with their praise for Kissinger. Senate Majority Leader Mike Mansfield, who had labored in vain for years to pass a bill forcing Nixon to bring the troops home, said, "I thought you did a magnificent job."

"Aren't you nice!" said Kissinger.

"Is the president in Saigon the roadblock?"

"No," said Kissinger.[16]

McGeorge Bundy, national security adviser to JFK and LBJ, said, "I think what you've done so far as it goes is magnificent. I just hope the next few steps turn out to be workable."

"Well, you know, we are dealing with two maniacal Vietnamese parties."

"I know it," said Bundy.[17]

At a press conference, Sen. Hubert Humphrey tossed aside a question about the settlement's effect on Democrats with words Nixon could only welcome: "Any loss of votes is better than a loss of life." If the president could make peace, Humphrey said, "he'll have my praise and my thanks."[18]

The so-called liberal news media, in the form of James "Scotty" Reston, *New York Times* columnist and primus inter pundits, glowed: "Well, your country owes you a great debt, Henry."

"Well, you're a good friend," Kissinger replied; "it's really meant a lot to me."[19]

Someone at the Associated Press dug up Kissinger's old *Foreign Affairs* article from 1969 and noted the resemblance between his proposals then and the settlement he'd described.[20] This did not help him extinguish talk of a coalition. The article described "partition"— Communists and anti-Communists governing their own parts of the South, as they would under a ceasefire-in-place—as a form of coalition government.

> A formal ceasefire is likely to predetermine the ultimate settlement and tend toward partition. Ceasefire is thus not so much a step toward a final settlement as a form of it.
>
> This is even more true of another staple of the Viet Nam debate: the notion of a coalition government. Of course, there are two meanings of the term: as a means of legitimizing partition, indeed as a disguise for continuing the civil war; or as a "true" coalition government attempting to govern the whole country. In the first case, a coalition government would be a facade with non-communist and communist ministries in effect governing their own parts of the country. This is what happened in Laos, where each party in the "coalition government" wound up with its own armed forces and its own territorial administration. The central government did not exercise any truly national functions. Each side carried on its own business—including civil war.[21]

The AP didn't mention that Kissinger had once described the division of the South into Communist and Saigon-controlled areas as "a disguise for continuing the civil war," not peace.[22]

"A Little Bit Diabolically"

Saigon tried to assert itself without breaking with the White House. After Kissinger's press conference, Thieu said there would be no peace in Vietnam until he signed a treaty himself.[1] His foreign minister issued a statement declaring that the South would never accept a settlement that worked against its people's "interests and aspirations"—without saying he meant the one that Kissinger had said was at hand. The *New York*

Times tried to interpret his words but missed the mark, concluding that the "statement appeared to be designed to emphasize that the main obstacle to an accord and a ceasefire was President Nguyen Van Thieu's doubts about what the details of the political settlement would be, not any general objection."[2] South Vietnam's president understood the details better than anyone at the *Times*, since he had read the settlement.

Thieu clarified his meaning the next day at a public rally of his National Assembly: North Vietnamese withdrawal was his "minimum demand." The *Times* sought further interpretation: "Informed Vietnamese and American observers here see President Thieu's continuing public objections to a ceasefire in place more as a delaying tactic rather than an attempt to impose a veto."[3]

Without South Vietnam's president on board, Nixon let Hanoi's October 31 signing deadline come and go. He dictated a statement to his press secretary that made a virtue of political necessity: "'After a long and difficult war, we're not going to have a hasty peace that might—that'll lead to another war—to the resumption of the war.' You get that line?"

"Yes, sir."[4]

As the press secretary walked out of the Oval Office, in walked Kissinger with Gov. Nelson Rockefeller, Nixon's old rival come to pay tribute. "You're still the strong man we've got to have," Rockefeller told Nixon. The governor had recently had dinner with managing editor A. M. Rosenthal and others on the editorial staff of the *Times*, who praised Kissinger's press conference performance. "They thought what he said was a hundred percent," Rockefeller said. "And then they wrote a good editorial."[5] This was putting it mildly; the *Times* had extolled the "truly remarkable achievement for the tireless professor-turned-diplomat," asserting that Kissinger "deserves the thanks of the nation and has certainly earned the respect of even the severest critics of the policies he has so doggedly pursued." And the *Times* got editorially huffy with Saigon: "Mr. Thieu has no grounds for complaint. The proposed agreement leaves his government intact, free to work out its own arrangements with the other side under the only kind of Vietnamization that ever made sense—Vietnamization of the peace."[6]

Later that day Rockefeller would tell the National Press Club that "history may well record that President Nixon has conducted the most successful four-year foreign policy of this century."[7] Kissinger gave Nixon the highlights. "In the speech he said, 'Now on Vietnam, the president's near success,'" said Kissinger. "'He's got the outlines of a peace agreement. He insists that all the details must be worked out because he doesn't want an armistice, he wants a peace. And can we, as Americans, be proud to have such a president?' Everybody got up and applauded."

"Did they?"

"Yeah, and that's the Press Club." Nixon was in "really the best position of all now, because the Right sees you are defending the details of the agreement. I'm now getting out the word that we want troops to leave," Kissinger said. "I triggered it with that *Los Angeles Times* story." He had misled reporter Bob Toth. "I put it out a little bit diabolically," said Kissinger. "I put it out as if we wanted all troops to leave."

The result was on the front page: "U.S. Insists Hanoi Pull Back Troops: Requires Action on Withdrawal of 145,000 Men before Signing." Reporters were calling the national security adviser to see if the *L.A. Times* story was true. "So that now when I say what we really want," Kissinger told the president, "people think it's moderate."

Vietnam was "going to come out in the best possible way," said Kissinger. It was overriding Watergate and dominating the news. It had voters focusing on "your strength, foreign policy," Kissinger said, "and all the other crap is dogs yapping at your heels."

"That's right," said the president.

"And it shows you as strong because you're not taking the easy way out of signing—"

"I like that. I like that."

"—just before election," said Kissinger. "By your not signing it, I think your moral position is absolutely unassailable." The adviser was looking past the president's anticipated landslide reelection. "I think we ought to start moving the B-52s further north," said Kissinger.

"Absolutely."

"Because the only thing those sons of bitches—and then, the day after your election, we ought to start reconnaissance around Hanoi again."

"Sure."

"And then, if after a week they don't answer, we ought to start bombing up there."

"They'll answer," said the president. "They know that after November 7th, if they continue the war, we'll knock the hell out of them." He asked how Hanoi's protest of his refusal to sign the deal by the October 31 deadline was playing in the press.

"Oh, boys will be boys," said Kissinger. He'd been working on reporters, downplaying the angry words flowing out of both North and South Vietnam: "I said, 'If I listened to everything that's said now from Hanoi or Saigon, we'd go out of our mind.' I said, 'Thieu is going to yell, Hanoi's going to yell.'" After the election, Kissinger said, they should send Hanoi "a seemingly very friendly letter saying now that you are reelected you want to reaffirm everything in the agreement. Now let's get it finished."

The president agreed. "I think, too, right after the election is over, we have to write Thieu a note telling him to get the hell on board here," Nixon said.

"Well, we've got Thieu quieted down now," Kissinger said.

"You really think so?"

"Yeah."[8]

The front page of the next day's *New York Times* suggested otherwise: "Thieu Calls Draft Accord 'Surrender to Communists': President, in a National Day Address, Denounces the Agreement as 'Only a Ceasefire to Sell Out Vietnam.'"

It was the most public possible blowup. National Day was South Vietnam's sorry version of Independence Day. It memorialized the November 1, 1963, overthrow of Ngo Dinh Diem, South Vietnam's first cruel but ineffective anti-Communist president, by a cabal of comparably cruel and ineffective anti-Communist generals (and then-Colonel Thieu) in a coup plot that climaxed with Diem's assassination following his surrender. Thieu chose his words carefully. Branding the settlement as a sellout and a surrender cut to the heart of Nixon's reelection campaign. The South Vietnamese president didn't call the election commission a coalition government, instead denouncing it as "a dictatorial three-part regime." He focused on the key strategic fact that Nixon and Kissinger had downplayed—the continued presence of 150,000 (Thieu said 300,000 to 400,000) North Vietnamese troops in the South.[9] The blowup was big: Thieu's speech led the evening news on CBS and ABC.[10]

Half of the national Democratic ticket seized the moment. Just as his brother-in-law Ted Kennedy had earlier assailed Nixon's strategy as fraudulent and failing, so now did vice presidential nominee Sargent Shriver. "I don't see what's the difference between what he has got and what he used to call surrender," Shriver said on San Francisco radio the day before Thieu erupted. It was the sharpest attack any Democrat had made on the settlement. "Most Americans were under the impression we were fighting for a free and independent South Vietnam," he said. "The total effort would be a failure if the government of Thieu fell."[11]

Shriver raised an astonishingly astute question in a speech the next day before the World Affairs Council: "What would happen, I keep asking myself, if [the agreement] were signed—if in fact, after we get out of there war should break out again. Is President Nixon or Dr. Kissinger, for example, promising Hanoi that we will not renew or participate in renewed fighting?" The highly classified answer was that for over a year the White House had been secretly assuring the Communists that Nixon would *not* intervene as long as Hanoi waited a year or two before conquering South Vietnam.

Shriver also asked, "Is he promising Saigon that we will return to fighting if the proposed agreement breaks down?"[12] Indeed, he was. The spot-on speculation continued when Shriver spoke to Protestant, Catholic, and Jewish clergymen in Baltimore the following day: "We might be promising Hanoi that we will not come back in and simultaneously promise Saigon that we would."[13]

Not one word of this made the evening news or the front page. Despite his Kennedy cachet, Shriver drew smaller crowds than Nixon, McGovern, or even Agnew. As his staff admitted, "We're number four."[14]

McGovern's reaction mattered much more, and he didn't attack the settlement. He attacked Thieu and Nixon for not signing it.[15] In a nationally televised address, speaking behind a lectern at First Methodist Church in Grand Rapids, Michigan, McGovern once again expressed his fervent doubt that Nixon would withdraw from Vietnam: "If he escapes his responsibility now, can you think he will end the war after the election, once he is free from the will of the American people?"[16] Reporters found out that Shriver had spoken out against the pact without consulting the top of his ticket.[17] After telling campaign audiences that the deal "won't work," Shriver was forced to add that McGovern wanted Nixon to sign it anyway, since it "would at least achieve United States withdrawal."[18] The Democratic message was a mixed, muddled mess.

The White House, however, wove Thieu's blowup into the narrative it had been constructing for four years. The *Times* cited "American officials" who told the newspaper that "Thieu is speaking to his domestic political audience only and does not pose as serious an obstacle to the realization of an accord as his words suggest." It was a preemptive countercharge—that Thieu's statements (rather than Nixon and Kissinger's) were designed merely for domestic political consumption.[19] But if you believed that Nixon was the kind of man who would just not let Saigon fall to the Communists—as the presidential nominees of both parties claimed—then the counterclaim sounded more true than the truth. The *Los Angeles Times* editorialized that Thieu was "digging in his heels in what appears a strategy to assert a better claim to national leadership."[20] The *Chicago Tribune* put the same opinion in a news article on the blowup: "It was to stifle 'internal' opponents."[21] A *Wall Street Journal* article likewise took White House officials at their word: "Yet many officials here believe this is largely jockeying for improved political and propaganda positions."²² Jockeying by Thieu, that is—not by the White House. Even things Kissinger was admitting to reporters, such as the continued post-settlement presence of North Vietnamese troops on Southern soil, sounded false when they came from Saigon. "President Thieu of South Vietnam insists, however, that the agreement allows Northern troops to remain in the South—but this cannot be taken at its face value, any more than his other claims," wrote the columnist Victor Zorza in the *Washington Post*.²³

It mattered little that Thieu's domestic political adversaries, including his opponent in the bitter, rigged 1971 election, Gen. Duong Van "Big" Minh, had united with him in opposition to the deal. Big Minh said, "As a military man, I am not for a ceasefire in place."²⁴ He wanted both sides to withdraw to regrouping areas, as they had in 1954. Yet in this rare case where Big Minh supported Thieu, the *Washington Post*'s reporter perceived veiled criticism beneath the surface: "Among those who are suggesting that perhaps Thieu is holding on to a position beyond the point of reasonableness is Gen. Duong Van (Big) Minh, who issued his annual National Day statement today."²⁵ It was at the annual luncheon Big Minh held honoring National Day that he announced that he, like Thieu, opposed the settlement.²⁶ When the facts don't fit a compelling narrative, the facts suffer.

The Chennault Affair

Why did Thieu's blowup in 1972 have so little impact when his blowup in 1968 had decided the presidential election? In both cases, Nixon controlled the narrative. He carefully prepared (or spun) the voters in advance to view the bombing halt as a gimmick. On October 25, 1968, the Republican nominee had issued a masterful misstatement of the facts:

> In the last 36 hours I have been advised of a flurry of meetings in the White House and elsewhere on Vietnam. I am told that officials in the administration have been driving very hard for an agreement on a bombing halt, accompanied possibly by a ceasefire, in the immediate future. I have since learned these reports are true.
>
> I am also told that this spurt of activity is a cynical, last-minute attempt by President Johnson to salvage the candidacy of Mr. Humphrey. This I do not believe.¹

While casting himself as the defender of Johnson's honor, Nixon did all he could to destroy the president's credibility. The fact was that Johnson had set three conditions for a bombing halt back in June and stuck with them until Hanoi accepted them all. It was only in October that the North finally accepted Johnson's demands that it (1) respect the DMZ, (2) sit down with the South in Paris, and (3) stop shelling South Vietnamese civilians. Nixon knew about these conditions because Johnson had briefed him throughout the negotiations. LBJ wasn't driving hard for a bombing halt. He knew how bad it would look to have one right before the election, and he fretted about how history would treat him. But even the hawks among his advisers told him they couldn't defend a decision *not* to halt the bombing after Hanoi accepted his demands.

Once Thieu "blew up," boycotting the Paris talks and tarring them as a step in the direction of a coalition government, it looked like Nixon's "cynical, last-minute" charge was true. What the voters didn't see was the behind-the-scenes effort Nixon had made to *get* Thieu to boycott the Paris talks in the first place. Using a prominent Republican fund-raiser named Anna Chennault as his go-between, Nixon secretly encouraged Saigon's refusal. Publicly, Nixon said he would do nothing to interfere with the peace negotiations; behind closed doors, he interfered. In 1968, Thieu's blowup perfectly fit the story the Nixon campaign was telling the voters about Johnsonian political manipulation. It was a deceptive story, but Nixon, through his own political manipulation, made it *look* true.

In 1972, Thieu's blowup didn't fit the story told by either nominee. That was the difference.

"The Clearest Choice"

In the final weekend of the campaign, McGovern assumed prophetic mode to issue "one more warning. If Mr. Nixon is elected on Tuesday, we may very well have four more years of war in Southeast Asia." To a *New York Times* reporter, McGovern sounded more bitter than ever. He proved once more that he had no idea what he was up against, saying of Nixon: "He has no plan for ending this war. He has not let go of General Thieu. He's not going to let that corrupt Thieu regime in Saigon collapse." McGovern was sure of that, even if Thieu wasn't. "He's going to keep the bombers flying. He's going to confine our prisoners to their cells in Hanoi for whatever time it takes to keep his friend General Thieu in office." McGovern was as certain as he was wrong. Self-righteous and self-defeating to the end, the prairie populist turned on those who would not heed his warning of four more years of war: "It's all right for people to be fooled once as they were in 1968. If they do it again, if they let this man lead them down the false hope of peace once again in 1972, then the people have nobody to blame but themselves."[1]

In fairness, a share of the blame belonged to the people's political leadership. When South Vietnam handed the Democrats a potentially devastating charge—that all the president had to show for four years of war and 20,000 American deaths was a deal that merely delayed and disguised defeat—the titular leader of the opposition scorned the very idea that Richard Nixon would ever abandon South Vietnam.

Nixon, for his part, did not use the last speech of his last campaign for office to warn or rebuke the people. The setting of the Election Eve campaign ad was simple and uncharacteristically casual: Nixon perched on the edge of a desk, bookended between framed family pictures, a simple curtain for a backdrop. Where McGovern seemed desperate, frantic, and shrill, the incumbent exhibited calm, confidence, and respect for the people's judgment:

> I am not going to insult your intelligence tonight or impose upon your time by rehashing all the issues of the campaign or making any last-minute charges against our opponents. You know what the issues are. You know that this is a choice which is probably the clearest choice between the candidates for President ever presented to the American people in this century.
>
> I would, however, urge you to have in mind tomorrow one overriding issue, and that is the issue of peace—peace in Vietnam and peace in the world at large for a generation to come.

He had done all in his power to obscure the real choice voters faced: between two paths to defeat, one straight and clear, the other concealed from American eyes, though revealed to the Communists in classified negotiations.

"As you know, we have made a breakthrough in the negotiations which will lead to peace in Vietnam," said the president. He ran through some of the popular provisions in the agreement with Hanoi.

He urged everyone to vote, whether for him or not.[2] But he offered his supporters a share in his glory, an opportunity to view themselves as he led them to view him—as people whose strength, steadfastness, and resolve led in the end to success in war and peace:

> There are still some details that I am insisting be worked out and nailed down because I want this not to be a temporary peace. I want, and I know you want, it to be a lasting peace. But I can say to you with complete confidence tonight that we will soon reach agreement on all the issues and bring this long and difficult war to an end.

You can help achieve that goal. By your votes, you can send a message to those with whom we are negotiating, and to the leaders of the world, that you back the President of the United States in his insistence that we in the United States seek peace with honor and never peace with surrender.[3]

Election Day 1972

As the votes came in, the president felt a strange sense of foreboding. "Melancholy," he called it, a word Lincoln would use. He was at a loss to explain it.[1] He was winning the greatest landslide in history, the networks were saying, though it would fall a bit short in the final tally.

At the "Western White House" in San Clemente the day before the voting, Nixon took a walk on the beach and spotted a peace sign some hippie had carved into a sandstone cliff. In the years since, it had nearly faded away. Nixon told his diary: "It looked like a man with a frown on his face. This may be an indication that those who have held up this sign finally have had their comeuppance and they are really in for some heavy depression." If he thought of any other way to interpret a vanishing symbol of peace, he didn't confide it to his diary. Instead, he suggested that God had a hand in his political success: "When I think of the ups and downs through the years, and particularly in this last year, I must say that someone must have been walking with us."[2]

A half hour before the *Spirit of '76* returned him to the nation's capital, Nixon summoned America's premier campaign journalist to the front of the plane for an interview. *The Making of the President 1972* would be a return to form for the author Theodore H. White, that form being the political campaign as heroic epic. Nixon spoke of his reelection as a national realignment. But something was a little off. He used an odd, film-noirish term for what he had done—"shifted allegiances"—a phrase that sounded too much like the "shifting allegiances" associated with turncoats, renegades, and betrayal. "Just think of the shift in the South," said Nixon. "You know what did it? Patriotism." A little defensively, he insisted that he wasn't "shooting blanks" regarding the promised peace in Vietnam: "I never shoot blanks, not when it comes to dealing with China or the Soviet Union."[3] None of these verbal twitches set off alarm bells for White. He was going to tell the kind of story he loved, one of political triumph and personal rebirth. The man before him was no longer the Nixon of old. White had seen it in the faces of the voters at his last campaign rally: "'No one loves Richard Nixon' had been one of the dominant clichés of American politics for years; but this crowd loved Richard Nixon, as did millions of others." As the interview ended and the *Spirit of '76* descended, the reporter looked out the cabin window and saw "the luminescence of Washington, all its floodlit shrines showing, glowing, disappearing, dissolving to runway lights."

The White House staff was waiting on the South Lawn for the president's helicopter. When he emerged, they cheered. The White House photographer snapped some shots of him triumphant.

After dinner, the president secluded himself in the Lincoln Sitting Room with his fireplace and his notepads.[4] He would get the election results by phone, alone. Something odd had happened: a cap had snapped off a top front tooth, one that had been part of his trademark grin for twenty-five years. A dentist had rushed to the White House and fitted the president with a temporary replacement. It hurt. And it might fall out, the dentist warned him, if the president smiled too broadly on television later. This did not improve his mood.[5]

Alone again, Nixon put on some music—the soaring score to *Victory at Sea*, the epic 1950s television documentary on the American Navy's World War II triumphs, including some in Asia.[6] It was composed by Richard Rodgers, with all the uplift and spirit he gave to the music of *Oklahoma, South Pacific,* and *The Sound of Music*.

The election results were reassuring. Historic. Richard Nixon won 60.7 percent (47,167,319 votes) to 37.5 percent for George McGovern (29,168,509 votes). Second only to LBJ's 61.1 percent in 1964. Nixon's margin of victory—17,998,819 votes—was bigger. The greatest in American history. He won every state except Massachusetts, 521 electoral votes in all. Second only to FDR, who won 523 in the first of his reelection campaigns. The South swung to Nixon hard. Six Dixie states gave him 70 percent of their votes or more. He shook other pillars of FDR's old coalition as well, taking 55 percent of the blue-collar vote, becoming the first Republican to win a majority of Catholics, and gaining the support of twice as many Jewish voters as he had in 1968.[7]

The president kept himself from grinning and losing his makeshift cap when he went on TV for a victory speech from the Oval Office at six minutes to midnight. At two points he started to break a smile, then stopped himself. He thanked his supporters and expressed "respect" for McGovern's: "I know that after a campaign, when one loses, he can feel very, very low, and"—the president smiled a little smile—"his supporters as well may feel that way." Back to a straight face. "And when he wins, as you will note when I get over to the Shoreham"—and again the president started to smile as he mentioned the hotel where Republicans were partying the night away—"people are feeling very much better."

The president put his own little twist on a line of poetry much of his generation knew by heart:

For when the One Great Scorer comes
To mark against your name,
He writes—not that you won or lost—
But how you played the game.[8]

The moral Nixon drew in victory was different: "The important thing in our process, however, is to play the game." His moral was oddly amoral. By this standard, however, Nixon had done the important thing. He played what he referred to in his victory speech as "the great game of politics."[9]

It was the first speech in which he brought himself to mention McGovern by name. "You know, this fellow to the last was a prick," Nixon told Kissinger long after midnight. "Did you see his concession statement?"

McGovern had been unable or unwilling to conceal his contempt: "Now the question is to what standards does the loyal opposition now rally? We do not rally to the support of policies that we deplore. But we do love this country and we will continue to beckon it to a higher standard." He invoked the words of the prophet Isaiah: they would rise with wings as eagles.[10]

"He was ungenerous," Kissinger told Nixon.

"Yeah."

"He was petulant."

"Yeah."

"Unworthy."

"Right." Nixon lamented the congressional returns. Republicans had picked up only twelve House seats, and in the Senate, astonishingly, lost two.[11] "All these left-wing columnists can do now is to piss on the not winning [of] the Senate and the House and building the party, but they couldn't care less about that. The main thing is, they know that we came up to bat against their candidate and beat the hell out of him."

"And came up against their issue and turned it into an asset," said Kissinger.

"That's right," said Nixon. "Don't you think so? Don't you feel that?"

"You made Vietnam your issue," said Kissinger.[12]

And yet through all the congratulations and celebrations, Nixon felt this muffled dread. In his memoirs, he would systematically list the possible reasons. The congressional returns? To some extent. Watergate? Maybe. "The tooth episode"? Probably. He also listed "the fact that we had not yet been able to end the war in Vietnam." The "yet" implied that at some point he did end the war.

Kissinger assured him that the election was "a tremendous personal triumph, Mr. President."[13] In a sense, Kissinger was right. The Richard Nixon that America acclaimed on this night was the one he had created out of words and polls. The one on their television sets who said that South Vietnam's right to self-government was non-negotiable and who got the Communists to agree to a settlement calling for free elections in the South. Who made them feel like heroes for standing by him as he continued to wage an unpopular war, saying it took four years to train and strengthen South Vietnam enough to stand on its own. Who reduced the number of American soldiers in Vietnam from over a half million to under 50,000 in four years without losing the war. Who insisted that he would fulfill his promise of "peace with honor" and that it would be "a lasting peace." Who made the election a referendum on patriotism and morality, as he defined the terms. The Nixon of Oval Office addresses and campaign commercials—the one that the man sitting before the fireplace in the White House had constructed through sheer force of will and disciplined self-control—had won in a landslide. They voted for the man Richard Nixon wanted to be.

They did not vote for the man he was. Not the Richard Nixon caught on his own secret tapes. Not the one who acknowledged behind closed doors that, even after four years of training for the South's soldiers, Saigon probably would never survive without American troops. Who kept Americans fighting and dying in Vietnam into his fourth year in office so Saigon wouldn't fall before Election Day and take his shot at a second term down with it. Who used Vietnamization as an excuse to prolong a war no one knew how to win. Who claimed success for it at politically opportune times—when he made one of his popular (partial) troop withdrawal announcements, and when he brought the last combat brigade home the month he was nominated for reelection. Who got Hanoi to accept his settlement terms by secretly assuring its Soviet and Chinese sponsors that it didn't really have to abide by them—that it could take over South Vietnam without fear of American intervention as long as it waited a "decent interval" of a year or two. Who sacrificed American lives for political gain—an act profoundly unpatriotic and immoral.

His strategy had won the election, not the war. Nixon had reason to feel foreboding on Election Night 1972. Voters had rejected all the things their president warned would happen if they voted for the Democrat: Peace with surrender. A bloodbath in Vietnam. The overthrow of the South Vietnamese government. Communist victory. All that was yet to come.

Promises and Threats

Nixon spent the next three months trying to get Saigon to shut up and take the deal.

This required extreme measures. South Vietnam rejected the settlement for the same reason that North Vietnam accepted it—because it would lead to Communist victory, as Nixon and Kissinger themselves admitted in candid, private moments.

Ordinary diplomatic pressure wasn't going to convince the Saigon government to accept a settlement that would lead to its destruction. The day after his landslide, Nixon sent his "get the hell on board" letter to Thieu: "Your continuing distortions of the agreement and attacks upon it are unfair and self-defeating" and "can only undercut our mutual objectives and benefit the enemy." With the venom came some meager carrots. While Nixon and Kissinger "consider the agreement to be sound," they would seek further concessions. One they'd already requested: the "*de facto* unilateral withdrawal of some North Vietnamese divisions in the northern part of your country." Another would say North and South Vietnamese soldiers "should be demobilized on a one-to-one basis" and return to their homes. (So much for the notion that the North's troops would "wither away" from the South thanks to settlement language about closing the Ho Chi Minh Trail.) They would clarify that the election commission wasn't a government at all and remove the embarrassing reference to there being just "three Indochinese countries."[1] They would ask for language saying that both sides would respect the DMZ.

In short, they would make the agreement look better without doing anything that would enable Saigon to survive it.

For Thieu to continue on his present course "would play into the hands of the enemy," wrote Nixon.[2]

On November 20, North Vietnamese negotiators returned once more to the white stucco villa in Gif-sur-Yvette to work on what President Nixon called "details." Saigon wanted sixty-nine changes to the draft, and Kissinger presented them all—including mutual withdrawal of American "and of all other non-South Vietnamese forces."

Le Duc Tho laughed: "So these are the technical changes, detailed changes, changes of the details?"[3]

The North Vietnamese still refused even to acknowledge the presence of their armed forces in the South. "These are not changes of details—these are not technical changes but these are political substantive changes. In consequence we will never accept them," said Le Duc Tho on November 21. Nixon and Kissinger had expected this. On technical issues, details, and rephrasing of the Vietnamese translation, the North Vietnamese showed willingness to negotiate. They accepted some of the proposed changes, none of them substantive. But they reintroduced a major substantive demand: the release of 30,000 civilian prisoners held by Saigon on suspicion of being Communist supporters.[4] The November round of negotiations ended with the two sides farther apart than in October.

Before leaving Paris, Kissinger read the North Vietnamese a cable he'd received from Nixon: "Under the circumstances, unless the other side shows the same willingness to be reasonable that we are showing, I am directing you to discontinue the talks and we shall then have to resume military activity until the other side is ready to negotiate."[5] Although framed as an order from the president, the cable arrived in Kissinger's hands marked "not a directive."[6] It was written for the purpose of threatening Hanoi.

Back in Washington, Nixon and Kissinger tried to resolve the negotiating impasse by threatening its source: Saigon. They asked that Nguyen Phu Duc (who was, basically, the South Vietnamese equivalent of Kissinger) be dispatched to the White House for a briefing on Nixon's "final position." Thieu agreed.

Before the two of them confronted Saigon's emissary in the Oval Office, Kissinger tried to explain away "that interview by that Italian bitch Fallaci." *L'Europa* had published an interview by Oriani Fallaci with quotations by Kissinger that seemed to confirm widespread suspicions about Kissinger's ego—and Nixon-specific suspicions that his subordinate built himself up with the press at the expense of the president. "I have always acted alone," Kissinger was quoted as saying. "The Americans love this immensely. The Americans love the cowboy who leads the convoy, alone on his horse; the cowboy who comes into town all alone on his horse, and nothing else."[7] In the Oval Office, Kissinger insisted that he had compared Nixon, not himself, to Gary Cooper in *High Noon*. As for himself: "I've never been on a horse in my life."

Haig escorted Duc in, and Nixon threatened South Vietnam's life. "I've met these past couple of days with the top leaders of the Senate and the House. Not the doves, the hawks. The tough ones," said the president. Hawks were the conservative Republican and Democratic supporters of the war. If Thieu didn't take the deal, Nixon said, the hawks would act when Congress convened in January. "Within ten days the Congress will cut off all aid. Economic and military. There's no question about it. The reason they will do that is not because they are for the Communists," said Nixon. It was because they were for the settlement he had negotiated with Hanoi. According to Nixon, the hawks unanimously agreed that the settlement reached in October met his goals of a ceasefire, return of the POWs, and, for the South Vietnamese, "the opportunity to determine their own future without having a Communist government imposed upon them against their will." Staunch supporters of the war—House Minority Leader Gerald R. Ford, R-Michigan, Sen. John C. Stennis, D-Mississippi, and Sen. Barry M. Goldwater, R-Arizona, the 1964 Republican presidential nominee Nixon called "Mr. Conservative"—would *lead* the fight to cut off aid when Congress convened, Nixon warned. "These people are friends of [South] Vietnam," Nixon said. "I don't want all your people to have died, 55,000 Americans to have died, for nothing. But on the other hand, I'm simply saying that if the Congress cuts off the pursestrings, we've got no choice."

This was 1972, the year Francis Ford Coppola's *The Godfather* entered American movie theaters and introduced a new line into the popular lexicon: "I'm going to make him an offer he can't refuse." Unlike the Mafia don played by Marlon Brando, the president didn't resort to euphemisms that veiled his meaning. Nixon was explicit: "Without aid, you can't survive. Understand?"[8]

Nixon had seen this kind of threat work before on Saigon. He'd taken part in making it. Lyndon Johnson had forced him to after the 1968 election. The day that LBJ found out that Republicans were secretly encouraging the South Vietnamese to boycott the Paris talks, he told the Tuesday Luncheon (his informal war cabinet) that "Nixon will double cross them after November 5." Three days after the election, Johnson forced Nixon's hand, subtly threatening to expose the Chennault Affair ("Now I don't want to say that to the country, because that's not good") unless the president-elect joined with the president in issuing an ultimatum to Saigon. The threat in 1968, as in 1972, was a loss of support in Congress. "You won't have ten men in the Senate support South Vietnam when you come in if these people refuse to go the conference," Johnson said.[9] In 1968, the ultimatum produced immediate results. The circumstances were different then: Johnson and Nixon were just demanding that South Vietnam sit down and talk with the North. In 1972, Nixon was demanding that it accept a settlement that its leaders considered suicide.

The day after Nixon's Oval Office ultimatum, Duc said to him, "It would be a choice either to die right now or to make a deal which would make us die."[10] The South refused to accept slow death even when threatened with the fast kind.

Nixon hadn't yet found a way to make the cutoff threat credible. Would congressional hawks really end American aid to Saigon? It would mean taking responsibility for losing the war.

Christmas Bombing

The Christmas Bombing of 1972 "was the most difficult decision concerning Vietnam that I made during my entire presidency," Nixon wrote of the order he issued in December 1972 dispatching B-52s to bomb Hanoi and Haiphong. "But I had no choice. I was convinced that if we did not compel the North Vietnamese to agree to our terms, Congress would force us to accept defeat by agreeing to a withdrawal in exchange for our POWs."[1]

Nixon's claims crumble under exposure to his declassified documents and tapes.

The decision to send in the B-52s was, for Nixon, easy. Back in October, after Hanoi *agreed* to his demands, he confided to Haig his intention to "bomb the bejeezus out of them."[2] A week before the election, he and Kissinger discussed hitting Hanoi with B-52s just to get a response to their request for another round of negotiations.[3] The day before it, Nixon told Kissinger, "After the election, we'll bomb the bastards."[4]

The B-52 attacks on Hanoi, in fact, were an obvious extension of Nixon's popular, poll-tested, and poll-confirmed bombing and mining of the North. He had used them to make Hanoi's acceptance of the "decent interval" terms look like military victory rather than surrender on the installment plan. He'd use the B-52s to do it once more.

Gov. Ronald Reagan independently came up with the idea of sending the B-52s north. When Kissinger informed him on December 15 simply that the bombing and mining were resuming, Reagan said, "I'd put the biggest fleet of B-52s they'd ever seen and just blow the hell out of the section of North Vietnam and just keep eyeballing them, making no more concessions and let this happen day after day until finally they recognize the only thing to do is put down the paper and sign this thing."

"Well, that's what we are doing," Kissinger said.[5]

Nixon's mind was made up to surge the bombing before the election, when the North was still clamoring to settle the war on his terms. Hanoi didn't have to give him a reason to bomb, just an excuse.

Hanoi provided one when it stepped up its demands—a response to the stepped-up demands Nixon and Kissinger made on Saigon's behalf.

When negotiations reconvened in December, the North was still willing to settle on the terms of October—the substantive terms that Nixon and Kissinger had been seeking for more than a year.[6] Le Duc Tho withdrew the demand for the release of 30,000 civilian prisoners by the South.[7] On December 12, he said he had to return to Hanoi for consultations—from there he would exchange messages with Kissinger to work out their remaining differences.[8] "So, you will be able to celebrate Christmas with your family," Kissinger said before they recessed on December 13.[9]

The next morning in the Oval Office, Nixon told Kissinger how they would sell the B-52 strikes on Hanoi to the American people. "What are the points we want to pound into the consciousness of these dumb, left-wing enemies of ours in the press?" the president asked rhetorically. "The enemy has never changed. The election didn't change it. The only friends we've got, Henry, are a few people of rather modest education out in this country, and thank God they're about 61 percent of the people, who support us." Kissinger would tell the press what Nixon wanted the 61 percent to believe. "I know everybody thinks they're dummies. They were smart enough to vote for us," said Nixon.

"Mr. President, they saved us. They're the good Americans," said Kissinger.

"But they've got to hear it clear and loud and simple. Prisoners, they will understand," said the president. Once again, they would portray military action as a way to free American POWs rather than as what it was: the reason for their continued imprisonment.

"While we know it's a sort of a . . . [*chuckles*] cynical thing, we're going to say, 'We're doing this because they won't return our prisoners before Christmas,'" Nixon had told Haig the night before. "They were supposed to do it, they wouldn't agree, and we're going to bomb until we get those prisoners back." The North had agreed in October to release the POWs; it hadn't backed away from that.[10]

Privately, Nixon acknowledged that Hanoi was the only party *not* backing away from the October agreement. Haig shared the president's reasoning with Kissinger by cable the day the negotiations recessed: "The American people would not understand and the realities were that it was the U.S. and not Hanoi that was backing away from the agreement because we had, in effect, placed additional demands on them. He also added that the other culpable party was Saigon and not Hanoi and that we can expect a massive push from the left charging us with being tools of Thieu."[11] (Nixon told a different story in *RN*: "It was vitally important that we lay responsibility for the current impasse where it belonged—squarely on the North Vietnamese.")[12]

Operation Linebacker II distracted America from Nixon's real problem: an inability to convince Saigon that the settlement offer was one it could not refuse. The distraction did real damage, however, both intentionally and not. Three thousand sorties over eleven days (skipping over Christmas Day) dropped 40,000 tons of bombs on the Hanoi-Haiphong complex, the most populous area in the North. Targets included power plants, shipyards, docks, railway yards, supply lines, and Radio Hanoi.[13] The veteran journalist Stanley Karnow, surveying the damage after the bombing, concluded that the B-52s "pinpointed their targets with extraordinary precision. Nevertheless, some bombs did stray, with ghastly results." More than 200 civilians died because they lived in the neighborhood of a railway yard. Another bomb meant for an airfield destroyed a wing of a hospital. A doctor told Karnow that he had had to amputate limbs to free patients from the rubble. The North Vietnamese said 1,318 civilians died in Hanoi, 305 in Haiphong.[14] Forty-two American crewmen died.[15]

For what? In 2013's *Henry Kissinger: The Complete Memoirs E-book Boxed Set,* Kissinger lists the concessions wrested from Hanoi between Election Day 1972 and the final agreement reached in the month after the Christmas bombing:

1. "In the negotiations since November 20 a number of changes had been achieved. The provision for our continued military support for Saigon had been expanded to permit in effect unrestricted military assistance." But Kissinger had already written that on October 19, 1972, Hanoi agreed to a "replacement provision [that], coupled with the massive augmentation of South Vietnamese forces even then under way, permitted what amounted to unlimited American military assistance to Saigon."[16] In other words, Hanoi had conceded this before the election, not after it. Kissinger is double-counting.

2. "The phrase 'administrative structure' to describe the National Council of National Reconciliation and Concord had been dropped, underlining its essential impotence." This was a purely verbal change in the Vietnamese translation for the guaranteed-to-deadlock election commission.

3. "The functions of the council had been further reduced, by taking away from it any role in 'the maintenance of the cease-fire and the preservation of peace' that had been in the earlier draft." Eroding the functions of a nonfunctioning-by-design commission does not constitute a substantive achievement.

4. "The Demilitarized Zone was explicitly reaffirmed in the precise terms of the provisions that established it in the Geneva Accords." These

were the precise terms that Hanoi had disregarded for years. Of course, Hanoi knew it could disregard the entire agreement without fear of American intervention if it waited a "decent interval."

5. "A provision had been added that the parties undertook to refrain from using Cambodia and Laos 'to encroach on the sovereignty and security of one another and of other countries.' This provision, aimed at the establishment of sanctuaries, was intended to reinforce the earlier one requiring the withdrawal of foreign forces."[17] Kissinger is double-counting again. During his October 26, 1972, "Peace Is at Hand" press conference, Kissinger announced that the parties had already agreed "to refrain from using the territory of Cambodia and the territory of Laos to encroach on the sovereignty and security of other countries."[18] (Moreover, Nixon and Kissinger had acknowledged privately to each other in October that whatever the North said, it wasn't going to withdraw from the Ho Chi Minh Trail through Laos and Cambodia.)[19]

6. "The international control machinery, now expanded to 1,160 people, was ready to begin operating on the day the agreement was signed."[20] Kissinger privately assessed the international control machinery in the draft agreement for Nixon on October 12, 1972: "We have more pages on international control, all of which is bullshit, to tell you the truth, but it will read good for the soft-hearts, for the soft-heads."[21] Kissinger did not claim to have improved the international control machinery since October, only to have expanded it and hastened the day it would begin operating. Since the international control machinery would oversee the North's paper withdrawal from Laos and Cambodia, Kissinger's private barnyard assessment is apt.

7. "A number of invidious references to the United States had been eliminated; a few additional technical improvements were made." Not even Kissinger claims these changes were substantive. (One does wonder how many "invidious references to the United States" had previously slipped past Nixon and Kissinger.)

8. "All the protocols and understandings essential to implementing the agreement effectively were completed."[22] The key understanding between the White House and the Communists, one Nixon and Kissinger would always deny, was the "decent interval."

That's Kissinger's entire list. People died for that.

The Joint Chiefs of Staff and the War in Vietnam, 1971–1973, a study by the Office of Joint History, asked how Linebacker II succeeded, since "the bombing campaign had been conducted in ways that, accumulated experience seemed to suggest, would surely fail." The military blamed civilian micromanagement for the ineffectiveness of the Rolling Thunder bombing campaign during the Johnson administration, but, according to the study's authors, Nixon micromanaged far more than his predecessor, intruding into details of target selection. "Perhaps Linebacker II succeeded because it ignored conventional wisdom and so surprised everyone, the North Vietnamese most of all," the study concluded.

Perhaps it worked because it could hardly have failed. Its goal was to persuade Hanoi to settle on terms that were substantively the same, and only cosmetically different, from the ones it had accepted before the election. Nevertheless, the Office of Joint History overstated the case in concluding that "Linebacker II clearly achieved its immediate objective. The North Vietnamese moved promptly toward an agreement and Thieu felt reassured enough to concur."[23]

The South Vietnamese president did not feel reassured, nor did he concur. Neither the Christmas bombing nor the final postbombing agreement did the one thing that would have allowed Saigon to survive: remove the North Vietnamese from the South permanently. (They didn't even do it temporarily.) The South Vietnamese remained unwilling to accept the slow death of a "decent interval" for as long as Nixon failed to convince them that their only alternative was sudden death by cutoff of American aid.

It was hard to get Saigon to take the threat of an aid cutoff seriously when Nixon could not produce any evidence that Congress was even considering an aid cutoff.

Nixon and Kissinger tried to make much of a couple of votes by the Democratic Caucus of the House and the Senate. On January 2, 1973, the day before the Ninety-Third Congress convened, House Democrats voted 154–75 to cut off funds for the war—but only under certain conditions. Those conditions were somewhat stricter than the ones their most recent presidential nominee had ridden to defeat. Unlike McGovern, who assumed based on past experience that Hanoi would release the POWs if he stopped the war, the House Democratic caucus insisted on "arrangements necessary to assure" the safe return of the POWs as well as the withdrawal of American troops.[24] A couple of days later, the Senate Democratic caucus took a similar position in a 36–12 vote.[25]

The votes in favor didn't amount to a majority in either house—and supporters would need two-thirds majorities in both to force the president to make a withdrawal-for-prisoners deal. Nixon could just veto it. The 75 House Democrats who were already on record against the idea would provide more than half the votes needed to sustain a veto. Nixon would need, at most, 71 more votes to put him over the top, and the incoming House included 192 Republicans. In other words, Nixon didn't even need a majority of his own party to kill a withdrawal-for-prisoners bill. (The notion that a congressional cutoff threat was truly looming at this time would remain credible to both hawks who hated it and doves who yearned for it, a triumph of myth over math.)

Truth be told, Nixon was *not* worried that Congress would force him to trade withdrawal for the POWs. He told Kissinger as much on December 14, the day he decided on the Christmas bombing. For one thing, Hanoi was not ready to make a simple swap of POWs for American withdrawal when a more comprehensive agreement was within its reach. Without Hanoi's agreement, "arrangements necessary to assure" the POWs' safe release could not be made.

President Nixon: Now . . . if we put it to the Congress straight out on the basis—and that's why I wanted to be sure that at one phase of it you had a specific case in there where it was POWs for withdrawal, which they would turn down—[if] we put it that way, we can bomb them for 10 years and have support from the Congress for Vietnam and for the bombing—*provided* it is not totally clear, I mean, that Thieu is to blame. That is the thing. Now on the Congress—in other words, we've got a lot more stroke with the Congress than I want [Thieu's adviser Nguyen Phu] Duc to know. As you know, we were playing that.[26]

Nixon's argument that he needed American troops in Vietnam to free the POWs had worked for years, though it was the opposite of true. The argument that he needed to bomb Hanoi to free the POWs would work just as well, though it was just as false.

Further reducing the threat of a cutoff was the unwillingness of the leading doves to take any action.

Sen. J. William Fulbright, D-Arkansas, famous for chairing the Senate Foreign Relations Committee's televised hearings on Vietnam in 1966 and for writing *The Arrogance of Power*, said his committee would not even take up the matter of a cutoff until after Inauguration Day—and only if Nixon had not reached an agreement by then. If there was no settlement by then, however, Fulbright said the committee would try once more to legislate an end to the war. But not for the time being. The committee met the day before the Ninety-Third Congress began. "The consensus of everyone present was that we did not wish to do anything to prejudice the negotiations," said Fulbright, in deference to power.[27]

Senate Majority Leader Mike Mansfield, the leader of so many efforts to stop the war, said this one had to wait: "I can see no way by which action could be taken between now and the inauguration."[28] Nothing could happen until Fulbright's committee voted on it first, and that would take several weeks at least, said Mansfield.

Inauguration Day would come and go without a settlement (Saigon the sole holdout) and without action by doves in either house of Congress. In the wake of devastating political defeat, it was a time for prudence among antiwar lawmakers.

As for the cutoff threat that Nixon was trying to brandish against Saigon—a total loss of American aid to the South Vietnamese government—Democratic doves weren't even *discussing* that in January 1973. One of them had just been hammered as the candidate of peace with surrender. An aid cutoff at that point would have been surrender without peace. What was the appeal of that? The Democratic caucus votes put them on record in favor of popular things—bringing all the troops and prisoners home safely—without so much as mentioning any unpopular thing that might have to be done or threatened in order to make the popular things happen.

Nixon's private assessment of the congressional doves was correct—they couldn't force him to make a withdrawal-for-POWs deal. At that point, they didn't even try.

The threat Nixon had made to Nguyen Phu Duc in the Oval Office after the election was failing to materialize. The president had been very specific about the timing: "Within ten days the [Ninety-Third] Congress will cut off all aid. Economic and military. There's no question about it."[29] Ten days after convening on January 3, 1973, Congress had not even begun to discuss a cutoff of aid to Saigon. This trend continued until Nixon got an idea.

He could start the discussion himself. He had told Saigon that hawks like Barry Goldwater, John Stennis, and Gerald Ford would lead the charge to cut off aid. On January 18, the president called Kissinger and asked if it would be useful to have Goldwater "say, 'Look, come along, boy'?"

"I think that might do some good," said Kissinger.

"I think if Goldwater could just come out and say it's time to quit this nonsense," Nixon said, "stop all this jabbering and this and that and—"

"And say that the major concern is now to close ranks," Kissinger added.

"I don't want one of the left to do it, but somebody like Goldwater from the right should say it."

"Exactly," said Kissinger.

"And maybe Stennis will say it if he won't," Nixon said. "Stennis should be another good one." Stennis was a Democratic hawk. That would make the threat bipartisan. That had worked for Johnson after the 1968 election, when the Democratic president and the Republican president-elect had joined together and (secretly) threatened Saigon with an aid cutoff if it didn't take part in the Paris peace talks.

Nixon and Kissinger would conceal their part in the making of this threat. "None of them would say they had talked to me," said Kissinger.

"Yeah," Nixon said, "well, you can tell them that it's very important that this not appear to come from the White House."

"Right."[30]

Goldwater immediately agreed.

"The difference is between them and us," Kissinger told him. "I mean, we shouldn't say that, but just for your information—cannot be explained to the American people."

"No, that's for sure," said Goldwater.[31]

Minutes after hanging up, Kissinger called Goldwater back. "Barry, the only thing I wanted to add is, you won't say that you and I talked," Kissinger said.

"Oh, hell, no," Goldwater said. "No, it's all on me."

"Good."[32]

Nixon had chosen the perfect mediums for this message. It got just enough attention, but not too much. It got a subhead on page one of the *New York Times:* "Goldwater and Stennis Tell Saigon Not to Balk."

"Two of the Saigon government's strongest supporters in the United States Senate—Barry Goldwater and John C. Stennis—urged the government today not to create obstacles to a peace agreement between the United States and North Vietnam," the *Times* reported. "They warned that South Vietnam would lose support in the United States for further economic and military assistance if President Nguyen Van Thieu blocked a settlement of the war."

"The South Vietnamese will need economic and military aid in the coming years," Stennis said on the floor of the Senate. "However, the South Vietnamese can jeopardize American support for such programs if they emerge now as the obstacle to peace in Southeast Asia."

Goldwater issued a statement warning Saigon not to object to what he called minor points: "It would imperil any future help which South Vietnam might obtain from this country."[33]

The aid cutoff was not the top story. It wasn't even the top story about Vietnam. The banner headline in the *Times* was, "U.S. and North Vietnam Announce Negotiators Will Meet Tuesday to Complete Text of Peace Pact."

All three TV networks on January 18 led with the story that peace appeared near. (Kissinger and Le Duc Tho had, in fact, completed the draft agreement five days earlier.)³⁴ None of the television news broadcasts mentioned Goldwater, Stennis, or the threat of an American aid cutoff at all.³⁵

On the morning of Inauguration Day 1973, the president had his national security adviser draft one more ultimatum to Thieu: The South Vietnamese leader had to accept the settlement by noon the next day or Nixon would tell congressional leaders he'd authorized Kissinger to initial the pact without Saigon. "If his answer is that he will not concur in the initialing of the agreement, [then] the congressional leaders in my view without question then will move to cut off assistance," Nixon told Kissinger. "I don't know whether the threat goes too far or not, but I'd do any damn thing, that is, or to cut off his head if necessary."³⁶ (In his second inaugural address, the president would proclaim, "The peace we seek in the world is not the flimsy peace which is merely an interlude between wars, but a peace which can endure for generations to come.")³⁷

In the letter Kissinger drafted under Nixon's name, he cited the threats the two of them had secretly generated in the Senate: "It is obvious that we face a situation of most extreme gravity when long-time friends of South Vietnam such as Senators Goldwater and Stennis, on whom we have relied for four years to carry our programs of assistance through the Congress, make public declarations that a refusal by your government of reasonable peace terms would make it impossible to continue aid."³⁸

The bipartisan threat from the hawks had its desired effect. Nguyen Tien Hung, an adviser to Thieu, and Jerrold L. Schecter described the impact on the South Vietnamese president in *The Palace File*:

> In his memorandums to Thieu, Hung had cited Senator Goldwater as the most ardent supporter of South Vietnam. Hung had told Thieu how he had assisted Warren Nutter in advising Goldwater on Vietnam policy during the 1963 [sic] presidential campaign. For Nixon to use Goldwater in the letter as evidence that support against Thieu was building up in Congress had a strong influence on Thieu. He realized that even his staunchest supporters had deserted him.³⁹
>
> Finally, Thieu relented.⁴⁰

He had little choice. His alternatives appeared to be either a settlement that would destroy South Vietnam in time or an aid cutoff that would destroy it in no time. Saigon would die a violent death sooner or later. All else held equal, Thieu preferred later.

"Let Us Be Proud"

January 23, 1973, 10:01 p.m., the Oval Office. "Good evening. I have asked for this radio and television time tonight for the purpose of announcing that we today have concluded an agreement to end the war and bring peace with honor in Vietnam." Henry Kissinger and Le Duc Tho had initialed it in Paris that afternoon; the formal signing by Secretary of State William Rogers would take place on Saturday the 27th. "Within 60 days from this Saturday, all Americans held prisoners of war throughout Indochina will be released," said the president. "During the same 60-day period, all American forces will be withdrawn from South Vietnam. The people of South Vietnam have been guaranteed the right to determine their own future, without outside interference."

The Agreement on Ending the War and Restoring Peace in Vietnam was negotiated "in the closest consultation with President Thieu," the president said. "This settlement meets the goals and has the full support of President Thieu and the government of the Republic of [South] Vietnam."

He called for scrupulous adherence to the Paris Peace Accords: "We shall expect the other parties to do everything it requires of them."

To South Vietnam, he said: "You have won the precious right to determine your own future, and you have developed the strength to defend that right." To the North: "Let us build a peace of reconciliation."

To America: "Now that we have achieved an honorable agreement, let us be proud that America did not settle for a peace that would have betrayed our allies, that would have abandoned our prisoners of war, or that would have ended the war for us but would have continued the war for the 50 million people of Indochina."

To the wives and children and parents of the POWs: "When others called on us to settle on any terms, you had the courage to stand for the right kind of peace so that those who died and those who suffered would not have died and suffered in vain, and so that where this generation knew war, the next generation would know peace."[1]

The Paris Accords lifted President Nixon back to the highest level of public approval he had ever enjoyed, 68 percent in Gallup, a peak he had reached only once before, in the aftermath of the Silent Majority speech.[2] Before the announcement, public opinion had been split about the Christmas bombing—46 percent in favor, 45 against.[3] Afterward, it appeared to 57 percent of those polled that the bombing had helped bring about the settlement.[4] A drag on his popularity for a short season, the bombing had no lasting impact on Nixon's approval rating, which remained high until Watergate revelations caused it to dissolve.[5] Pollsters didn't even ask whether people thought the Goldwater/Stennis aid-cutoff threat helped bring about the settlement, although it had a bigger impact on Thieu than all the B-52 raids.

The ceasefire-in-place between North and South didn't quite happen. "Fighting Goes On; Saigon Reports Slight Decline in Level of Military Activity," the *New York Times* reported on February 1, five days after the signing.[6] The Vietnamese death toll for the first week of peace with honor was a staggering 4,295.[7] By the end of February, Saigon put it at 10,000.[8]

In March, even while American soldiers and POWs were still returning from Vietnam, Nixon and Kissinger considered sending bombers back in to hit North Vietnam. They were afraid Saigon might fall without a "decent interval" of a year or two.

On March 14, 1973, Kissinger told the president: "My worry is that if we don't scare them off within the next month, then they're going to make the decision now to attack in the fall. The whole thing can't survive. If it lasts two years, Mr. President—"

"There's no problem. I agree with that," said Nixon.

"But," Kissinger said, "if we let ourselves get rolled there now—"

"This year? That's what I mean. We can't do it," said the president. "I'm out on a hell of a limb on this."

"Me, too."

"And the administration, as we all talked about the fact that we achieved the peace," said Nixon. "Even the December bombing, you see, becomes very much in question." The Christmas bombing looked successful when voters thought it had led to peace; it would look less so if Saigon fell before the next Christmas.

Later that day with Haldeman, the president made it plain that Saigon would be expendable in a couple of years: "Well, Henry's exactly right. We've got to do everything we can to see that it sticks for a while, but as far as a couple of years from now, nobody's going to give a goddamn what happens in Vietnam, not one damn degree."

"Yeah, but it'd be tough if it breaks this year," said Haldeman.

"We hope not," said Nixon. "They might have an offensive, for example, this year and . . . we'll see."[9]

By March 16, the president had a new election-oriented timetable for Saigon's survival. Although he could not run for president again, he was concerned about the congressional midterm elections of 1974: "What you and I've got to do, Henry, is just to work on these big games, the big plays, and that's what we're going to do, by golly," Nixon told Kissinger on March 16. "And, essential to this is not to let them, if we possibly can, do us in on Vietnam this year. We can't let that happen. And that's why we're going to just bomb the bejeezus [*unclear*]."

"After the summer of '74, that's a different story," said Kissinger.

"No, after the elections of '74," said Nixon.

"After the election of '74," Kissinger said. "In fact, if it's got to happen, the spring of '75 is better than the spring of '76."

The president concurred.

Having established the optimum time for South Vietnam's demise, the president told Kissinger that it would not take place. "I don't think it's going to happen," said Nixon. "I don't have all that lack of confidence in the South; I don't have all that confidence in the North, Henry."

"I don't think it's going to happen, Mr. President, if we pull them up short a few times."[10]

They didn't, however. Nixon didn't order any air raids on the North, and, after the last POWs returned at the end of March 1973, he couldn't—not without risking terrible consequences for himself.

The Prisoners Dilemma

In the months Nixon had spent trying to get Thieu to sign the Paris Accords, he sent letter after letter promising to "respond with full force should the settlement be violated by North Vietnam," to "insure that the provisions of a peace settlement are strictly enforced," to "take swift and severe retaliatory action."[1] South Vietnam's president didn't buy it. As an NSC aide reported to Kissinger on December 20, 1972, Thieu argued that "it would not be possible for the US to retaliate against North Vietnam if the Kissinger/Tho agreement were violated, since the US would not risk having new POWs."[2] Why would Thieu say this during the Christmas bombing, which gave Hanoi a chance to shoot down fifteen B-52s and capture twenty-four new prisoners?[3]

There was a world of difference between bombing the North before the Paris Accords and bombing it afterward. Before, Nixon had something to trade for the release of the POWs: total American withdrawal. The Accords, however, removed all American troops from Vietnam in return for the POWs. Under the Accords' terms, both the troops and the POWs exited Vietnam during February and March 1973. That meant that as of April 1, 1973, Nixon could not send American bombers back over North Vietnam without putting himself in a military, diplomatic, and political bind. What would he do when Hanoi shot down a plane and took its crew prisoner? How could he free them then? He couldn't count on a daring commando raid to rescue them. He'd tried that at Son Tay in 1970 and found an empty prison. He couldn't offer total withdrawal of American ground troops again; Hanoi already had that. No matter how he spun the Christmas bombing to the public, he knew he couldn't force Hanoi to release the POWs by bombing it more; that was how Hanoi *got* prisoners. Nixon could offer to stop bombing in return for the POWs' release, but then he'd be accused of surrendering. It wouldn't be the subtle, concealed surrender of a "decent interval," but the open, public, undeniable kind.

Or he could just keep bombing. If NSSM-1 was right, and Saigon couldn't survive without American troops on the ground, then Nixon would get the POWs back after the North conquered the South. But in that case, he'd have to share responsibility for military defeat. If NSSM-1 was wrong, and American airpower was enough to keep Saigon going—even without American advisers there to get South Vietnamese soldiers to hold their ground and create "lucrative targets" out of North Vietnamese troops, or to direct the bombing thereof—then the POWs would remain in Hanoi for as long as Hanoi kept fighting. If Nixon had bombed the North after April 1973, his options would shrink to surrender, defeat, or letting the POWs rot indefinitely.

That was the military and part of the political dilemma that bombing would have posed for him. The diplomatic dilemma had scandal potential. Nixon got the North Vietnamese to sign the Paris Accords in the first place by giving Moscow and Beijing secret assurances that they could violate it. Through Kissinger, he had let Hanoi's biggest allies know that as long as it waited a year or two before taking over the South, he would not intervene. In other words, he would *not* enforce the agreement. It was with this understanding that China and the Soviet Union had used their influence to encourage Hanoi to sign. Obviously, bombing the North would have risked the relationships Nixon had created with the Soviets and Chinese. Less obviously, it would have risked public exposure. He and Kissinger both assumed the Communists taped their closed-door meetings. If Nixon had intervened following the settlement, Moscow or Beijing could have responded by playing tapes of Kissinger telling them he wasn't going do that. A diplomatic problem would then have risked becoming an additional political debacle.

In February and March 1973, while the Paris Accords were still bringing the troops and POWs home, Nixon said in the Oval Office on more than one occasion that he would not be able to bomb the North. He even said it to a newly freed prisoner visiting him on March 12, Captain Jeremiah Denton. It came up in a discussion of reconstruction aid. Nixon was trying to enlist POW support for providing American aid to Hanoi, an idea that was growing more unpopular as the prisoners revealed more details of their suffering. "Aid is a tool that we can use," Nixon said. "We might restrain them. I'm not sure. They're savages, they're barbarians, they're terrible people, they're primitive and all that." But if they violated the agreement, America could cut off aid.

"If we don't have this tool, what other tool do we have? Well, we can threaten, look, we'll bomb you again," said the president. "But that's an idle threat. It's not totally idle, because they think I'm a little, shall we say, crazy."[4] Nixon didn't explain why the threat was idle—or why Hanoi would have to succumb to Madman Theory to take it seriously.[5] With British prime minister Edward R. G. Heath on February 1, Nixon made a similar point: "Henry's got them [the North Vietnamese] convinced there's a madman in the White House who doesn't care about polls or anything and might bomb you again."[6]

If sending the B-52s back in would be a self-destructive, even irrational, act for Nixon, failing to do so would be risky as well. It would leave him open to the charge of abandoning an ally, failing to keep a promise (Thieu could make Nixon's letters public, and eventually did), and snatching defeat from the jaws of victory. He'd convinced majorities that the May 8 and Christmas bombings succeeded; how could he justify not defending the South with American airpower?

The Final Cutoff

On June 29, 1973, Minority Leader Gerald "Jerry" Ford rose on the floor of the House of Representatives and made an announcement that left his colleagues baffled and disbelieving. The chamber's top Republican announced that President Nixon would sign a bill banning any and all American combat activity in, over, or off the shores of North Vietnam, Laos, Cambodia, and South Vietnam. To House Republicans and Democrats, conservatives and liberals, hawks and doves alike, Ford's announcement sounded preposterous.

Just two days earlier, Nixon had vetoed legislation to ban American bombing in Cambodia. The president had denounced the bill, saying it would "cripple or destroy the chances for an effective negotiated settlement in Cambodia and the withdrawal of North Vietnamese troops."[1]

Cambodia had its own powerful Communist insurgency by then, one of the many disastrous unintended consequences of Nixon's secret bombing four years earlier. When the bombing destabilized the neutralist regime of Prince Norodom Sihanouk, rightists in Phnom Penh ousted the prince in a 1970 coup. Sihanouk saw his only path back to power in an alliance with Cambodia's then-tiny group of Communist rebels, the Khmer Rouge. Sihanouk became the Reds' best recruiting tool in the Cambodian countryside, where the hereditary monarch commanded fanatical loyalty, and by 1973 the Khmer Rouge had grown in size and strength to the point where it threatened to take over the country. Kissinger was trying to negotiate a Cambodian ceasefire, and Nixon was using American bombers in an attempt to make the Communists settle. This time, his bombing of Cambodia was open, not secret.

House doves had tried to override Nixon's veto of the Cambodian bombing ban on June 27, but fell 35 votes short.[2] In the intervening two days, they had not found the votes they needed. "We did not have the votes to override the veto," Rep. Paul Whitten, R-Connecticut, admitted on June 29. "We do not have them now." So why was Nixon yielding to them just two days after proving he could defy them? If he had enough votes to keep bombing Cambodia, why would he agree to ban all military intervention there and in every other country in Indochina? It didn't appear to make any sense.

Neither did the other presidential concession Ford announced that day: "If military action is required in Southeast Asia after August 15, [1973, the day the bombing ban would take effect,] the president will ask congressional authority and will abide by the decision that is made by the House and the Senate, the Congress of the United States." With this, Nixon surrendered his constitutional advantage over congressional doves. Before, they needed to muster a two-thirds majority to defeat him; after August 15, they could do it with a simple, bare majority. What was going on?

Lawmakers on both sides told Ford he had to be mistaken. The House minority leader said he had checked with a White House spokesman the night before and again that morning. "I did not talk to the president, but I am talking of people who have told me they have talked with the president," said Ford. This was not good enough for his colleagues.

"I would not be willing to accept the word of some member of the White House hierarchy as to the views of the president," said Rep. Sidney Yates, R-Illinois. "Much too frequently such statements have later been repudiated as not correctly expressing the president's ideas."

"I am very nervous about the question of assurances from spokesmen in the White House," said Rep. Robert Giaimo, D-Connecticut. "There is one person who by a simple press statement today could clarify this matter, and yet we hear nothing but silence from the one man who could clear it: the president of the United States."

Ford left the floor of the House and placed a call to San Clemente, California, where the president was staying at his "Western White House." There is no known tape of this conversation, since there was no known secret recording system at the Western White House, but the White House Daily Diary shows that the conversation between Nixon and Ford took place. When the House minority leader returned to the floor, he said, "I just finished talking with the president himself for approximately 10 minutes, and he assured me personally that everything I said on the floor of the House is a commitment by him."

His colleagues were still baffled ("No President has ever signed a similar piece of legislation," said Rep. Paul Cronin, R-Massachusetts),[3] but they no longer disbelieved. On this day, many representatives who had supported Nixon on the war for years, many who had opposed him just as long, as well as many in the middle joined together and cut off all funds for American combat throughout the nations of Indochina by a vote of 278 to 124.

Not once during the debate did any lawmaker raise the possibility that the president would one day use this vote to blame them for losing Vietnam. Neither did anyone in the Senate, which voted for the all-Indochina combat ban that same day, 64–26. Liberals declared victory. Humphrey urged his colleagues to seize the opportunity, since "the fact is we do not have the votes to end this war without some agreement with the man in the White House." McGovern called it "the happiest day of my life," for "at long last, the administration has finally capitulated."

Stabbed in the Back

"We won the war in Vietnam, but we lost the peace," Nixon wrote in *No More Vietnams*. The Paris Accords depended on "a credible threat of renewed American bombing of North Vietnam." Congress, however, "destroyed our ability to enforce the peace agreement, through legislation prohibiting the use of American military power in Indochina."[1]

So why did he invite Congress to do just that? The explanations Nixon gave don't quite add up. In his memoirs, he wrote that "it seemed clear that another cutoff bill would be proposed and that I could not win these battles forever."[2]

It didn't seem that way to two of his most experienced vote counters, Ford and Mel Laird, the former defense secretary who had moved to the White House as counselor for domestic affairs. Laird had spent eight terms in the House representing Wisconsin; Ford had led all of Nixon's previous fights in the House against legislation that would "tie his hands" in Vietnam. These were men who knew the House, and they said that this was a battle Nixon could win.

The night before Nixon offered to let Congress "tie his hands," June 28, 1973, Kissinger asked Laird whether the administration had to accept a congressional ban on American military action in all of Indochina. The answer was no. Ford had told him the House would stand firm behind limiting the ban to Cambodia and would not impose it until August 15. Kissinger's secretaries transcribed this phone conversation.[3] It's clear that Laird and Ford thought Nixon had the votes he needed.

Doves had hoped to force Nixon to accept a Cambodian bombing ban with a hardball legislative move. They attached the ban as an amendment onto a continuing resolution, a measure that was needed to avoid a government shutdown since Congress hadn't passed all the necessary appropriations bills. The *Wall Street Journal* went so far as to call one such measure passed on June 26 "a practically veto-proof prohibition against continued U.S. bombing in Cambodia and Laos."[4] The *Journal* was wrong. Nixon vetoed the allegedly veto-proof measure, and Congress sustained him.[5]

Like Ford and Laird, the *New York Times* found that Nixon had the votes he needed to win: "There was a majority of votes in both Houses of Congress to attach amendments cutting off bombing funds to fiscal measures that the government needs to continue operating into the new fiscal year starting Sunday. But Congress did not have the two-thirds vote necessary to override presidential vetoes of these measures."[6]

While neither the vote-counting experts nor the votes themselves back up Nixon's claim that he had no choice, he does get support from an unlikely corner: the liberals who thought he would never let Saigon fall. Daniel Ellsberg explains:

> Mort Halperin has pointed out to me that without the challenge of Watergate hanging over him, Nixon could almost surely have mustered the one-third-plus-one votes he needed to defeat a congressional attempt to override his veto, in a situation in which he could claim to be "enforcing a signed agreement" by bombing.... But with the [Sen. Sam] Ervin [D-North Carolina] hearings [by the Senate Watergate Committee] approaching, and [former White House counsel John] Dean's testimony on Nixon's own obstruction of justice impending, Nixon could not afford to use up political capital peeling off votes against bombing when he would need every vote he could get to fight off impeachment. Therefore, in June Nixon reluctantly reached a deal

with both houses whereby all bombing would be ended on August 15. Probably most members of Congress thought of this as affecting only the bombing of Cambodia, which went on openly until that deadline.[7]

This explanation misreads both the politics and the timing. Nixon not only could sustain a veto after the Senate Watergate hearings and John Dean's televised testimony, he did. The hearings began in May 1973, Dean testified on June 25 and 26, Nixon vetoed the Cambodia-only bombing ban on June 27, and the House sustained his veto later that day. Also, contrary to Ellsberg's surmise, the House and Senate debate makes it clear that lawmakers knew the combat ban they voted for on June 29 applied to all of Indochina—it's why they were baffled at Nixon's suggestion. The biggest problem with the dove explanation, however, is the notion that the Vietnam issue would deplete Nixon's political capital. He had just ridden the issue to a landslide victory. He had beaten McGovern's withdrawal deadline legislation—and then beaten McGovern himself—by successfully charging that a vote for either was a vote for surrender and defeat. He could have made the same charge against any bill banning the use of American airpower in Vietnam and beaten it, too.

Instead, he chose not to fight a battle he could win. Why not?

"We Can Blame Them for the Whole Thing"

Unfortunately, there are no White House tapes from June 23 to July 8, 1973, when Nixon was in San Clemente, so they can't give us a direct answer. Earlier tapes, however, show that Nixon recognized the political value of letting Congress "tie his hands" in at least one country of Indochina.

"On Cambodia, we've got to bomb the goddamn place until the Congress takes away the power," Nixon told Kissinger on March 29. "We can blame them for the whole thing going to pot." Could the Cambodian government survive without American bombing?

"No," said Kissinger. "Cambodia cannot survive without it. It may not even survive *with* our bombing. In fact, I was wondering—"

"Maybe we just let her go. Make a deal now," said the president.[1]

If Cambodia fell while Nixon was still bombing there, he'd get the blame for losing it. But if it fell after Congress stopped the bombing, he could pin the blame on the legislature.

"We really have to think about whether we are not better off saying these sons of bitches just are responsible for the defeat," Kissinger told

Laird on June 25, 1973—that is, just four days before Nixon issued the invitation that allowed him to blame Congress for defeat not only in Cambodia, but in Vietnam.

"Politically, you'd be better off. I don't think Cambodia will ever work out very well anyway and I'd like to be able to blame these guys for doing it myself," said Laird. "Henry, you know I'm kind of a black character, and I'd like to blame these guys for the incapability of getting these things resolved, because I think it's damn touch-and-go to get it resolved as far as Cambodia is concerned anyway."

"Yeah," said Kissinger.

"I'd like to be able to blame them," said Laird.[2]

Nixon didn't need to save Cambodia from Communism; to profit politically, all he had to do was blame its loss on the Democrats. This was how Cold War politics was played. Republicans didn't come up with a way to save Eastern Europe from Communism after World War II, but blaming FDR and Truman for its loss helped the GOP win the House and Senate in 1946. No one came up with a way to save China from Mao and Zhou, but claiming the Democrats lost it helped Republicans gain House and Senate seats in 1950. A Republican president was unable to achieve the victory his party called for in Korea, settling instead for an armistice dividing the country, but Reagan found a way to blame Democrats for defeat in that war. Likewise, Democrat JFK never found a way to save Cuba from Castro, but blaming Republicans for its loss helped make him president. Time after time throughout the Cold War, blaming political opponents for losing countries to Communism proved to be a highly effective way of winning elections. Politically, it paid.

Nixon clearly saw the political benefits to be reaped by allowing Congress to bar American combat of any sort in, over, and off the shores of Cambodia (and let's not forget Laos). Could he have failed to see the political benefits of allowing Congress to "tie his hands" in North and South Vietnam as well?

With free hands, the president would face an unpalatable choice. The most likely outcome of either of his two alternatives—to bomb or not to bomb—was that Nixon himself would have gotten the blame for losing Vietnam. Both alternatives pointed to surrender or defeat. Either would have undone all his political accomplishments, destroyed the conservative coalition he had painstakingly constructed to replace FDR's, and carved the following epitaph into every obituary, encyclopedia entry, and textbook biographical box: "Richard M. Nixon was the first American president to lose a war."

As an alternative, he could extend his hands to Congress and tell Jerry Ford that it could go right ahead and tie them. This would solve his political problems. (It also prevented other, more thorny, political, diplomatic, and military problems from arising.) When Saigon fell, he could blame Congress.

He didn't even wait for it to fall. The day he signed the bill, August 4, 1973, the president wrote an open letter to the Democratic Speaker of the House and Senate majority leader castigating Congress for being so irresponsible as to accept his invitation:

> I would be remiss in my constitutional responsibilities if I did not warn of the hazards that lie in the path chosen by Congress.... With the passage of the congressional act, the incentive to negotiate a settlement in Cambodia has been undermined, and August 15 will accelerate this process.
>
> This abandonment of a friend will have a profound impact in other countries, such as Thailand, which have relied on the constancy and determination of the United States, and I want the Congress to be fully aware of the consequences of its action. For my part, I assure America's allies that this Administration will do everything permitted by Congressional action to achieve a lasting peace in Indochina....
>
> I can only hope that the North Vietnamese will not draw the erroneous conclusion from this Congressional action that they are free to launch a military offensive in other areas in Indochina. North Vietnam would be making a very dangerous error if it mistook the cessation of bombing in Cambodia for an invitation to fresh aggression or further violations of the Paris Agreements. The American people would respond to such aggression with appropriate action.[3]

"In congressional quarters," the *New York Times* reported, the letter "was widely interpreted as an attempt to shift onto Congress the blame and the responsibility if Cambodia should fall to the Communists after the halt in the bombing."[4] The liberals should have foreseen that Nixon would likewise blame Congress for the loss of Vietnam. Of course, they didn't believe he would ever allow Saigon to fall. For the remainder of his presidency, however, Nixon did not once ask Congress for authority to use American military force in North or South Vietnam.

In *Years of Upheaval*, the second volume of his three-part memoirs, Kissinger wrote that he objected to the decision to accept a ban on American combat in Indochina: "When I protested to Nixon, he said it was too late; he had yielded to *force majeure*," a legal term for uncontrollable events that excuse people from fulfilling their contractual obligations.⁵ *Force majeure* is French for "greater force," but Nixon had yielded to a *lesser* force, one he had the power to overcome. An explanation presents itself in the pattern of Nixon's secret political decision making. We know from Nixon's tapes that he timed the withdrawal of American ground troops to the 1972 election so he could avoid blame for losing Vietnam. We also know from his tapes and declassified documents that he had Kissinger negotiate a "decent interval" deal so he could avoid blame for losing Vietnam. We do not *know* from Nixon's tapes and documents that he invited a congressional ban on American combat throughout Indochina so he could avoid blame for losing Vietnam—but to do so would have followed the pattern he set. To do *otherwise* would have required a profound change of character.

Vote counters inside and outside of the White House said Nixon could beat back an all-Indochina combat ban.⁶ It was, however, to his political advantage simply to throw the fight. I think he did.

Nixon's *Dolchstoßlegende*

On December 7, 2006, a date which should live in infamy but probably won't, a *New York Times* article declared Richard Nixon's Vietnam War strategy a success. The *Times* was trying to place in historical perspective the front-page news of the day: a bipartisan panel had called for total withdrawal of American combat troops from one of the two wars that were then in the midst of eclipsing Vietnam as America's longest and least popular.¹ In a sidebar headed "Military Analysis," the paper's chief military correspondent, Michael R. Gordon, took aim at the panel's proposal to withdraw American combat troops from Iraq in fifteen months. The headline asked, "Will It Work on the Battlefield?" Gordon's answer was no. "The military recommendations issued yesterday by the Iraq Study Group are based more on hope than history," Gordon began. He provided this version of history: "It took four years, from 1969 to 1973, for the Nixon administration to make South Vietnamese forces strong enough to hold their own and withdraw American combat forces from Vietnam. Even so, when Congress withheld authority for American airstrikes in support of those forces in 1975, the North Vietnamese quickly defeated the South and reunified the country under Communist rule."²

Gordon's history lesson packs a lot of wrong into two sentences. It took four years for Nixon to hide his *failure* "to make South Vietnamese forces strong enough to hold their own." Nixon kept American troops fighting that long because he realized that Saigon would fall without them and that his reelection prospects depended on delaying Communist victory past November 1972. Moreover, President Ford never asked Congress for authority to launch air strikes in Vietnam, so it's just not true that "Congress withheld authority for American airstrikes in support of those forces in 1975." Worse, Gordon ignores the leading role that Nixon and Ford both played in denying the president the authority to order such strikes. Besides, no one has ever shown how Nixon or Ford could have solved the problems that sending in the B-52s would have caused.

The problem is not that one journalist got some facts wrong. All Gordon did was summarize the conventional story of how the war ended (while getting some facts wrong). The conventional story is the problem, because Richard Nixon authored it. Compare Gordon's history lesson to the one Nixon gave in *No More Vietnams:*

> All that we had achieved in twelve years of fighting was thrown away in a spasm of congressional irresponsibility.
>
> When the Paris peace accords were signed in January 1973, a balance of power existed in Indochina. South Vietnam was secure within the ceasefire lines. North Vietnam's leaders—who had not abandoned their plans for conquest—were deterred from renewing their aggression. Vietnamization had succeeded. But United States power was the linchpin holding the peace agreement together. Without a credible threat of renewed American bombing of North Vietnam, Hanoi would be sorely tempted to prepare to invade South Vietnam again. And without adequate American military and economic assistance, South Vietnam would lack the power to turn back yet another such invasion.
>
> Congress proceeded to snatch defeat from the jaws of victory. Once our troops were out of Vietnam, Congress initiated a total retreat from our commitments to the South Vietnamese people. First, it destroyed our ability to enforce the peace agreement through legislation prohibiting the use of American military power in Indochina. Then it undercut South Vietnam's ability to defend itself, by drastically reducing our military aid.[3]

As we've seen, declassified tapes and documents destroy the claims Nixon makes in the passage above. It's a stabbed-in-the-back myth, an American Dolchstoßlegende. Such myths have an obvious appeal to the guilty. They shift the blame for defeat from their authors to their critics. Yet if you take the excerpt from *No More Vietnams* quoted above, remove its scathing moral judgments, reduce its essentials to a two-sentence summary, and toss in a couple of inaccuracies of the kind that crop up under deadline pressure, then you've got the history lesson that appeared twenty-one years later in the *New York Times*.

None of this means that the *Times*'s chief military correspondent is some kind of partisan political hack. On the contrary, the myth Gordon retailed as history is, bizarrely, bipartisan, embraced not only by the authors of America's defeat in Vietnam but by their critics as well.

In the political debates over Vietnam in the 1970s, the Right and Left shared certain beliefs about Richard Nixon. Generally, both believed that Nixon would do all in his power to prevent Communist victory in Vietnam; specifically, both believed that Nixon would defend Saigon with American airpower after the troops came home. But the two ends of the political spectrum differed in their *attitude* toward their shared beliefs. The belief that Nixon would always defend South Vietnam with American airpower made the Right happy and the Left unhappy; the belief that Congress stopped Nixon made the Right unhappy and the Left happy. On their fundamental assumptions about Nixon's intentions, however, the Right and Left were in agreement.

The Right and Left were both wrong. Evidence has emerged in bulk since Nixon's death in 1994 ended his long, twilight struggle to keep Americans from finding out what was on his tapes and in his White House documents. But evidence on tape and paper sitting in archives unheard and unread can do nothing to shatter a belief. In 1999, for example, the National Archives had declassified the Nixon White House tapes from February through July 1971—the tapes referenced at the beginning of this book. In 2001, nevertheless, Larry Berman published *No Peace, No Honor: Nixon, Kissinger, and Betrayal in Vietnam*, in which he argued:

> The reality was the opposite of the decent interval hypothesis. . . . The record shows that the United States expected that the signed treaty would be immediately violated and that this would trigger a brutal military response. Permanent war (air war, not ground operations) at acceptable cost was what Nixon and Kissinger anticipated from the so-called peace agreement. They believed that the only way the American public would accept it was if there was a signed agreement. Nixon recognized that winning the peace, like the war, would be impossible to achieve, but he planned for indefinite stalemate by using the B-52s to prop up the government of South Vietnam until the end of his presidency.[4]

Berman was still telling readers in 2001 what McGovern told voters in the fall of 1972: a vote for Nixon was a vote for four more years of bombing. As for all of the available tape-recorded evidence that Nixon's real strategy was to postpone, not prevent, Communist victory, Berman didn't mention it. Any of it.

Upon such omissions rests the continued belief in Nixon's Vietnam Dolchstoßlegende in the twenty-first century. It's not that people don't believe the evidence on tape and in the declassified documents; it's just that, by and large, most people have never heard of it.

Nixon's backstabbing myth depends on suppression of evidence. The original suppressers were Nixon and Kissinger, who kept the taped and written evidence from the American people by taking advantage of the national security classification system. For them, suppression of evidence was a conscious, deliberate choice, a means of deception. For writers trying to understand Nixon and Kissinger, however, suppression of evidence can sometimes be an unconscious reflex. Confirmation bias is the psychological term for the mind's bad habit of reinforcing assumptions by seizing on evidence that supports them and by discounting, rejecting, or ignoring evidence that challenges them. Although abundant evidence has emerged to shatter the assumptions shared by the Right and Left about Nixon and Kissinger in the 1970s, confirmation bias tosses the evidence aside in favor of preserving the assumptions. Whether conscious or unconscious, deliberate or unintended, deceitful or naïve, suppression of evidence has permitted a deadly myth about the past to threaten our present.

Unearthing Nixon's Strategy

Jeffrey Kimball has done more than any other scholar to bring Nixon's "decent interval" exit strategy to light. The author of two landmark works on the subject, 1998's *Nixon's Vietnam War* and 2004's *The Vietnam War Files: Uncovering the Secret History of Nixon-Era Strategy*, Kimball did the essential, painstaking work of unearthing the evidence buried in the archives, analyzing it rigorously to determine which mysteries it solved and which ones remained impenetrable, and distilling his findings into books that meet the highest standards of scholarship. Many of the tapes and documents I've quoted in this book first appeared in *The Vietnam War Files*. Kimball's battle to overturn assumptions that have been held across the political spectrum for decades has been long, uphill, and ultimately successful—but only in one arena, academia. Among university historians who study Vietnam, Kimball's work commands respect.

Outside academia, his work is virtually unknown. While *Nixon's Vietnam War* got good reviews in some of the publications whose coverage determines whether readers at large will hear of a book's existence, *The Vietnam War Files* got no reviews at all in popular journalistic outlets. That's not unusual for a scholarly work, but in this case the oversight was tragic. The general American public at the start of the twenty-first century *needed* to know how Nixon really ended the Vietnam War—before politicians, pundits, and policy entrepreneurs started holding him up as an example to follow in Iraq and Afghanistan. Unfortunately, in this age of Internet-driven research, journalists who investigate Nixon's exit strategy are likely to find misleading, inaccurate information.

The Vietnam War Files fell victim to a particularly nasty academic hatchet job. Kimball once wrote a critical review in *Diplomatic History* of a young historian's first book, *A Bitter Peace: Washington, Hanoi and the Making of the Paris Agreement* by Pierre Asselin.[1] *Diplomatic History* subsequently let Asselin review Kimball.[2]

Asselin's method was to distort Kimball's work, then assail the distorted version. For example, Kimball provided more than enough evidence to finally settle the debate over whether Nixon embraced Madman Theory. As Kimball explained, Nixon "meant to convey his supposed madness as irrationality, unpredictability, unorthodoxy, reckless risk-taking obsession, and fury." Note that Kimball referred to Nixon's "*supposed* madness." Although Kimball wrote that the president "may indeed have had some sort of personality disorder" (a possibility that occurs to other readers of his transcribed conversations), the historian did not write that Nixon was actually mad. Asselin, however, falsely claimed that Kimball made this false claim: "That Nixon wanted to 'jar' and 'create fear' (pp. 59–60) in the Hanoi leadership to precipitate the end of the war does not constitute evidence of madness, as Kimball would have it." Kimball never said it did; in fact, he carefully laid out the *rational* basis for convincing one's adversary of one's madness. When Asselin complained of "the paucity of evidence for Nixon's alleged madness," the casual reader might have assumed that Kimball alleged it. But Kimball didn't. The rest of Asselin's review was just as aggressively misleading.[3]

Asselin did his worst where Kimball is at his best—on Nixon's "decent interval" exit strategy. Kimball was the first historian to discover Kissinger's scribbled note in the margins of the briefing book for his first, secret trip to China: "We want a decent interval. You have our assurance." Of this smoking gun, Asselin wrote: "That evidence, however, is far from definitive as there is no proof Kissinger actually told the Chinese everything he considered telling them, including the reference to a decent interval."[4] Asselin's point is half-trivial, half-ludicrous. Of course Kissinger didn't tell the Chinese everything he considered telling them.

Proof that Kissinger negotiated a "decent interval" on Nixon's behalf is, nevertheless, in the near-verbatim transcripts the NSC staff made of the conversations with Chinese and Soviet leaders. Kissinger, it is true, did not use the words "decent interval" in his first conversation with Zhou Enlai, but he laid out in detail the settlement terms that would produce such a delayed defeat: total American withdrawal, return of all American POWs, and a ceasefire-in-place for "18 months or some period." As Kissinger told Zhou: "If the government is as unpopular as you seem to think, then the quicker our forces are withdrawn, the quicker it will be overthrown. And if it is overthrown after we withdraw, we will not intervene."[5] Kissinger didn't need to use the words "decent interval" in order to negotiate one. In a later conversation with Zhou, Kissinger referred to the waiting period as a "reasonable interval," a "sufficient interval," and a "time interval"—phrases too close to "decent interval" to miss.[6] A fair reviewer would have mentioned the abundant evidence; Asselin didn't.[7] "Even if Kissinger did raise the decent-interval idea with the Chinese," Asselin wrote, "there is a distinct possibility that he did so not because he was sincere but because he viewed the idea as a 'carrot' to get Beijing to encourage Hanoi to reach an agreement."[8] By 2006, when Asselin's review appeared, there was more than enough taped evidence available to show that Kissinger really was negotiating a "decent interval" exit for Nixon, including the August 3, 1972, conversation in which Kissinger said that "we've got to find some formula that holds the thing together a year or two," and the October 6, 1972, conversation in which Kissinger said, "I also think that Thieu is right, that our terms will eventually destroy him." The National Archives released those last two conversations in December 2003—too late for Kimball's book, but three years before Asselin's review.

Not content with ignoring evidence, Asselin twisted some: "The questionable sincerity of Kissinger's statements to the Chinese is suggested in a memorandum of conversation reproduced in Kimball's work in which Zhou Enlai catches Kissinger contradicting himself (p. 192)."[9] The evidence on the page Asselin cited shows the opposite of what Asselin claimed.

> Zhou: You said that if a regime should be subverted by an outside force, then you would intervene.
> Kissinger: No.
> Zhou: Then there must have been a mistake in the record.[10]

Kissinger wasn't contradicting himself; he had told Zhou that if the North overthrew the South after Nixon withdrew the last American troops, Nixon would *not* intervene. The evidence that Asselin used to undermine Kimball's argument is actually further evidence that Kimball is right. It's yet another example of Kissinger negotiating a "decent interval."

The crowning absurdity is that Asselin himself acknowledged in *A Bitter Peace* that all Nixon and Kissinger managed to get was a "decent interval." Here's how Asselin summarized the dilemma Thieu faced in January 1973 after Nixon and Kissinger got Goldwater and Stennis to threaten South Vietnam with a cutoff of American aid:

> If he refused to sign the agreement, assistance from Washington would end soon. If he signed, he might thereby not only provide Nixon, Kissinger, and others with the "decent interval" they wanted between the American withdrawal and Saigon's collapse but encourage Congress to continue financial assistance to his government during the interval.... The second alternative was less unappealing than the first only because it bought a little time.[11]

Exactly. Thieu took the "decent interval" deal Nixon and Kissinger made because the alternative for Saigon was a quicker collapse. Asselin knew this—before the foremost historian of the "decent interval" gave *A Bitter Peace* a critical review.

Asselin's attacks crumble upon exposure to evidence. The problem is that few people are familiar with the evidence. In the age of Web-based research, anyone who looks for criticism of Kimball's work will quickly find Asselin's review. Since it was published in an academic journal by a university professor and has enough footnotes to reassure a reader who doesn't check them, Asselin's review offers the appearance of scholarship without the substance.

John Carland, a retired State Department historian, wrote, "The most convincing rebuttal to the ['decent interval'] theory, revolving around a typical set of claims about Nixon and Kissinger's supposed machinations, can be found in Pierre Asselin's 2006 review essay in *Diplomatic History* of Kimball's 2004 *The Vietnam War Files*."[12] Carland was in a position to know better, having edited two of State's massive compilations of historical documents on Nixon's Vietnam War strategy: *Foreign Relations of the United States (FRUS), 1969–1976: Vietnam, January–October 1972* and *Foreign Relations of the United States, 1969–1976: Vietnam, October 1972–January 1973*. The *FRUS* volumes published by State's Office of the Historian gather together hundreds of White House, Pentagon, CIA, State, and other agency documents, many of them formerly classified, into the official historical record of American foreign policy.

Being the government's official version of events, the *FRUS* volumes are enormously influential. New volumes are available free online, making them enormously convenient as well.[13]

Naturally, Kimball was disturbed when documents crucial to understanding Nixon and Vietnam were, in his words, "inexplicably absent" from Carland's *Vietnam, January–October 1972* volume, such as Kissinger's May 27, 1972, conversation with Gromyko ("We are prepared to leave so that a Communist victory is not excluded"); the August 3, 1972, tape about finding a formula to hold Saigon together for a year or two; the September 26, 1972, exchange between Kissinger and Le Duc Tho regarding "a suitable interval"; and more. These omissions were "unfortunate and troubling," wrote Kimball.[14]

The most unfortunate and troubling omission concerns the date that the volume ends, October 6, 1972. Carland reproduced a Kissinger-drafted letter from Nixon that day assuring Thieu that there was "no basic difference as to the objectives we both seek—the preservation of a non-Communist structure in South Vietnam which we have so patiently built together."[15] But Carland omitted any mention of Kissinger's tape-recorded admission to Nixon that same day that the settlement they were about to make would destroy South Vietnam.[16] The "our terms will eventually destroy him" tape provides crucial context; without it, readers may not see that Nixon and Kissinger's assurances to Thieu were hollow.

Carland closed the government's official historical record of the January–October 1972 period of Nixon's Vietnam War with his own interpretive note:

> Furthermore, Hanoi's proposal as structured reflected to a substantial degree Nixon and Kissinger's longstanding desire to separate the military and political issues in the negotiations. By separating the issues, the American leaders believed that the military ones could be agreed to quickly with relative ease and, in the wake of such agreement, the United States could honorably depart, leaving the political issues to be settled by the Vietnamese parties in further negotiations.[17]

Neither Nixon nor Kissinger nor Hanoi nor Saigon believed that "the political issues [would] be settled by the Vietnamese parties in further negotiations." Nixon and Kissinger's division of negotiating issues into "military" (American withdrawal, release of the POWs, a ceasefire-in-place for the North Vietnamese) and "political" (the question of who would govern South Vietnam) was euphemistic, a way of evading the reality, recognized by all the parties, that Hanoi's and Saigon's armies would resolve the "political" issue by fighting it out. Carland's *FRUS, 1969–1976: Vietnam, January–October 1972* is flawed by both the omission of essential documents and the commission of interpretative howlers. (In another "Editorial Note," Carland wrote, "President Nixon was certain he had presented a negotiating proposal that could produce a lasting peace." The claim that Nixon "was certain" of this depends on suppression of the evidence that he was not.)[18] Carland launched a broadside attack on Kimball, heavy with scorn but light on evidence, in *Passport: The Society for Historians of American Foreign Relations Review*. Calling "The Decent Interval Theory" an "argument for which the evidence used by proponents fails to support their interpretive rhetoric," Carland demanded additional evidence while failing to confront (or even acknowledge) the evidence at hand:

> Questions that have not been answered and backed by documents must be answered—with evidence. To wit: When did this decision "to pursue a decent interval solution" take place? What were the precise circumstances surrounding the decision? Were there pre-decisional meetings on this subject? Surely there had to be, so when did they occur? What was said by whom at these meetings? Where's the beef?[19]

In other words, if Nixon and Kissinger didn't put their decision to prolong a war and fake peace for political gain through their normal bureaucratic planning process—complete with "pre-decisional meetings" (because "surely there had to be")—then it just didn't happen. Carland's standard is absurd—not absurdly high, just absurd. Nixon and Kissinger could not afford to take the risk of putting their strategy through the bureaucracy, because the larger the number of people who knew about it, the greater the chance it would leak. Sacrificing the lives of American soldiers for the sake of a reelection campaign and negotiating a "decent interval" were, by political necessity, furtive activities. If Nixon and Kissinger hadn't kept them secret, they would have lost their jobs. In this regard, the decision to seek a "decent interval" was like the decision to launch the Watergate cover-up. Are we to assume that the cover-up didn't occur because we lack "the precise circumstances surrounding the decision" or documents from "pre-decisional meetings" detailing "what was said by whom"? (For that matter, are we to assume that the Watergate break-in did not occur, due to a similar lack of "pre-decisional" bureaucratic documentation?) The fact that we can hear Nixon on tape orchestrating the Watergate cover-up proves that at some point—we do not know exactly when—he decided to do so. The lack of an orderly, bureaucratic paper trail leading to these decisions is not surprising, given the normal human reluctance to produce self-incriminating evidence. The proof that Nixon and Kissinger timed military withdrawal to the 1972 election and negotiated a "decent interval" comes from extraordinarily rich and undeniable sources—the Nixon tapes and the near-verbatim transcripts that NSC aides made of negotiations with foreign leaders. In the face of the evidence, Carland just sneered:

> Scholars who argue that Kissinger—by telling Chinese and Soviet rulers that the United States had no intention of returning to Vietnam after a settlement had been reached and American forces had been withdrawn, and indeed was content to let historical forces work themselves out to create the future—somehow managed to convey, in Marxist-Leninist jargon, a long-in-place Nixon administration decent interval policy must have decoder rings the rest of us do not possess.[20]

Scholars don't argue that. Kissinger didn't merely say that the United States had no intention of returning after a settlement. He was shockingly specific about the length of time required between Nixon's final troop withdrawal and the South's final defeat. He used the same number with the Chinese and the Soviets: eighteen months.[21] And he warned that Nixon might intervene after a settlement if the Communists failed to give him that much time.[22] In other words, Kissinger informed the Communists that Nixon would not intervene if the North took over the South after a "decent interval," but that he might if they tried to do it too soon. No "Marxist-Leninist jargon" or "decoder ring" required.

The University of Virginia's Miller Center

I learned of Nixon and Kissinger's exit strategy the slow way. The University of Virginia's Miller Center has scholars listen to the secret tapes of Presidents Franklin Roosevelt, Truman, Eisenhower, Kennedy, Johnson, and Nixon, prepare accurate transcripts, and supply the necessary background so readers and listeners can get the most out of these priceless recordings. My first assignment after joining the Miller Center in 2000 was to start listening to the Nixon tapes from day one and write a summary of every Oval Office conversation for review by an editorial board of eminent historians. Before I moved on to a new assignment, I had heard every Oval Office conversation Nixon held from February 16, 1971, through April 15, 1971—in other words, the conversations quoted and discussed at the beginning of this book.

Listening to the Nixon presidency unfold hour by hour (suppression of evidence was not an option), I couldn't help but notice Nixon and Kissinger's "decent interval" exit strategy. At the start of the taping, Nixon sounded optimistic, firm, resolute. "We can lose an election, but we're not going to lose this war, Henry," he said on February 18, 1971, when the taping system was still new. But Kissinger revealed how little this meant when he summarized their negotiating strategy a few minutes later. Kissinger planned to meet with Le Duc Tho in the fall of 1971 and "tell him, 'Look, we're willing to give you a fixed deadline of total withdrawal next year for the release of all prisoners and a cease-fire.' What we can then tell the South Vietnamese—they've got a year without war to build up."[1] Already they realized that a settlement would not bring peace. It would just provide an interruption to the conflict, "a year without war," during which Nixon would withdraw the last American troops. This withdrawal would enable Nixon to deny responsibility for losing the war.

In reality, however, the withdrawal would lose the war. The tapes prove that Nixon and Kissinger realized South Vietnam would fall without American troops. Leaving meant losing. They concealed that from the American people, but it was the basis of their military and diplomatic strategy. It was the reason they stretched American military withdrawal out for four years. "We can't have it knocked over—brutally—to put it brutally—before the election," as Kissinger put it on March 19, 1971.[2] Almost two years before the election, Nixon had decided to bring the last troops home sometime between July 1972 and January 1973. Call it the window of opportunism. He preferred to do it a little before the election; Kissinger, a little bit after, but both were determined to keep American troops in Vietnam long enough to keep the Communists from winning until the election had passed.

Vietnamization was not a strategy Nixon seriously pursued; it was a fraud he perpetrated. He told America that Vietnamization was "a plan in which we will withdraw all of our forces from Vietnam on a schedule in accordance with our program, as the South Vietnamese become strong enough to defend their own freedom."[3] His tapes prove that he didn't believe this for most of the time he was claiming it in public. His actions speak even louder than his recorded words. By keeping American troops in Vietnam long enough to avoid a preelection defeat, Nixon demonstrated by deed as well as word that he expected Saigon to fall without them.

Out of context, some of Nixon's public statements about Vietnamization sounded more modest, as when he said it would give Saigon "a chance to avoid a Communist takeover."[4] Technically, a snowball has a chance in hell. In that sense, Nixon's program also had "a chance." But by that same standard, Saigon had "a chance" to survive on its own in 1971, 1970, 1969, 1968, 1967, 1966, and 1965, when the first American combat troops arrived. The president's slipperier public comments don't relieve him of responsibility for the unambiguous—and false—claims he made:

> In Vietnamization we have withdrawn our forces as rapidly as the South Vietnamese could compensate for our presence. But we have not withdrawn them so as to allow the North Vietnamese to impose a political future on the battlefield. . . .
>
> The fundamental question remains: can the South Vietnamese fully stand on their own against a determined enemy? We—and more importantly the South Vietnamese—are confident that they can.[5]

A postmodernist or a lawyer might be able to find ambiguities in these statements from Nixon's February 25, 1971, State of the World report. The words, nevertheless, have a plain meaning. Nixon put it this way on another occasion: "Just as soon as the South Vietnamese are able to defend the country without our assistance, we will be gone."[6] He made a clear public commitment: to keep American troops in South Vietnam until it could defend itself without them, and to bring them home once it could. The truth was that he would keep American troops in South Vietnam until he was sure it wouldn't fall before Election Day, because he did not believe Vietnamization would make Saigon capable of surviving once he brought them home. His public claims about Vietnamization were not just vague or ambiguous; they were dishonest.

Decision Points

Some of the things Nixon said in the privacy of the Oval Office, however, *are* ambiguous, even self-contradictory. I include the contradictions in this book for two reasons: they're relevant, and they reveal Nixon's character. Most people shy away from the darker implications of their choices, and Nixon was no exception. Sometimes, however, circumstances forced Nixon to face the consequences of his actions. Those circumstances arose when he had to make a decision about strategy. At these key decision points, Nixon and Kissinger spoke bluntly, even brutally, about the consequences.

After Nixon made a decision, however, he could say anything he liked about it. He could portray it and himself in heroic terms. He could engage in spin (for outsiders) and justification (for insiders). While he was making a decision, however, Nixon had to face some ugly truths.

The tapes capture four key decision points:

Nixon's Decision to Time the Final American Troop Withdrawal to the 1972 Election. Before the April 7, 1971, troop-withdrawal announcement, the choice facing Nixon was whether to announce that he would bring the last of the troops home from Vietnam by the end of 1971 (as a large majority of the public favored) or to adopt a secret timetable delaying their return long enough to keep Saigon from falling before Election Day.

In making this decision, Nixon revealed (if only to Kissinger) his real views on Vietnamization. Leaving shortly before the election (Nixon's preference) would have given Hanoi too little time to overthrow the South before Americans voted; leaving shortly after the election (Kissinger's preference) guaranteed Saigon's survival through the election. (Since Hanoi didn't agree to Nixon's terms until October 1972, and Saigon didn't accept them until January 1973, the final troop withdrawal took place in March 1973.)

Nixon's secret timetable was based on the realization that without American troops defending Saigon, the North would defeat the South (although "not easily"). The timetable would also create the appearance of success. Americans would see in the months leading up to the presidential election that (1) the number of American soldiers in Vietnam had decreased from more than 500,000 to fewer than 50,000, and (2) Saigon was not (yet) falling. This made Nixon's claim that Vietnamization was working seem plausible, even though he adopted the timetable based on his realization that Vietnamization would fail.

Yet on April 8, 1971, the morning after Nixon went on television and announced his decision to withdraw 100,000 soldiers from Vietnam by December 1971 (rather than bring home all 284,000), Nixon said to Kissinger, "You realize if we get out the way we want, we win it."

"That's right," said Kissinger.

"If the communists don't win, we win it," said Nixon. "That's what victory is in Vietnam, Henry. It was never to conquer North Vietnam. It was simply to save South Vietnam."

"That is right," said Kissinger.

"Right?"

"Right."[1]

The president who had decided to time withdrawal to his reelection realized he was not saving South Vietnam, just delaying its demise. He was saving his reelection campaign. But once he had made the decision to withdraw on a secret, election-based timetable, Nixon could cast his actions as heroic and count on Kissinger to cheer him on. Like most people, Nixon said things to make himself feel and look better.

Nixon's Decision to Propose a Ceasefire-in-Place. The choice facing Nixon before Kissinger's secret May 31, 1971, negotiating session with Hanoi was whether to explicitly drop their demand for "mutual withdrawal" and accept the continued presence of North Vietnamese troops in the South after the final American withdrawal by formally proposing a ceasefire-in-place.[2] Once again, Saigon's survival was at stake.

Mutual withdrawal would have meant the North Vietnamese withdrawing their troops when Nixon withdrew America's. This would have left the South Vietnamese army to handle just the Vietcong guerrillas, something that NSSM-1 said it could do (unless and until the North Vietnamese army moved back into the South). Hanoi, however, never agreed to withdraw from the South. Nixon couldn't force the issue diplomatically, since no one had come up with a military strategy that would force the North Vietnamese out of the South and keep them out.

At the May 1971 secret talks, Kissinger proposed unilateral American withdrawal coupled with a ceasefire-in-place for the North and South Vietnamese—an arrangement that, if NSSM-1 was correct, would lead to Saigon's downfall.

Two days before making this crucial concession, Nixon and Kissinger had a blunt and brutal discussion about Saigon's prospects. "If it's got to go to the Communists, it'd be better to have it happen in the first six months of the new term than have it go on and on and on," said Kissinger. "I'm being very cold-blooded about it."

"I know exactly what we're up to," said Nixon.[3]

Before deciding to propose a ceasefire-in-place, Nixon faced facts. After he made the decision, however, he lectured Kissinger on how tough and Teddy Rooseveltian he was.[4] Postdecisional self-puffery, however, couldn't undo the consequences of the decision.

Hanoi's Decision to Drop Its Demand for Thieu's Resignation. The North's August 1, 1972, concession was crucial to Nixon and Kissinger's exit strategy. Up to that point, Hanoi had not only insisted that the settlement replace Saigon's government with a three-party coalition government, but it had also demanded that Thieu himself play no part in the coalition. Thieu's ouster from power would have made a "decent interval" impossible, if American intelligence analysts were correct in their belief that the regime would quickly unravel without him. Nixon all along rejected Hanoi's demand to get rid of Thieu, saying it was the equivalent of asking him to overthrow the South Vietnamese government himself.

Hanoi's agreement to let Thieu remain in power at the time of the settlement meant that, after nearly four years of fruitless negotiations, a "decent interval" deal was finally coming within Nixon's reach. "For the first time," Nixon wrote in his memoirs, "the Communists actually seemed to be interested in reaching a settlement."[5]

Once again, Saigon's survival was at stake; once again, Nixon and Kissinger had a blunt, brutal discussion. "I think we could take, in my view, almost anything, frankly, that we can force on Thieu," Nixon said on August 3, 1972. "South Vietnam probably can never even survive anyway."[6]

"We've got to find some formula that holds the thing together a year or two," said Kissinger.[7]

This didn't stop Nixon from later telling Kissinger that Saigon had to survive, "period."[8]

This discussion raises another interpretive issue: When Nixon asks whether America could have a viable foreign policy if the North took over the South, and when he and Kissinger discuss whether it would worry the Chinese, it sounds like they're making a decision based, at least in part, on how it affects America in the world. But they'd already made the decision. They already *had* a formula that would hold South Vietnam together a year or two—the withdrawal of all American troops, a ceasefire-in-place for the two Vietnamese sides, and secret assurances to the Communists that Nixon would not intervene if they waited approximately eighteen months (i.e., a year or two) before the North took over the South. As Kissinger would remind Nixon at the next turning point, if Hanoi simply accepted his terms (as it was going to do), then he would have a deal. (The politically unacceptable alternative for Nixon would be to back away from his own settlement terms, which would raise fundamental questions about what was wrong with them.)

On August 3, 1972, Nixon and Kissinger were merely rationalizing a decision they had made long ago.

Nixon's Decision to Let Hanoi Settle on His Terms. Both Nixon and Kissinger boast in their memoirs that Hanoi finally accepted their settlement terms in October 1972.[9] Kissinger summarized the key provisions of Le Duc Tho's October 8, 1972, proposal: "Its essence was an Indochina-wide ceasefire, withdrawal of U.S. forces, release of prisoners, no further North Vietnamese infiltration, the right to resupply the South Vietnamese forces up to existing levels, and above all, the recognized continuation of the existing political structure in Saigon, eligible for U.S. economic and military aid—the basic program we had been offering since May 1971 and that had been consistently rejected."[10]

The no-infiltration provisions, as we have seen, were insubstantial, but the rest of Kissinger's summary is accurate. Two days before the pivotal October 8 negotiating round, Kissinger gave Nixon the blunt, brutal bottom line: Thieu said their terms would destroy South Vietnam, and he was probably right. Nixon stuck with his terms.

The last decision—in one way the smallest, since it just confirmed decisions Nixon had made previously—best illustrates the importance of decision points, as well as the relative *unimportance* of whatever Nixon said about the decision after he made it. From the October 8 negotiating round until the signing of the final agreement on January 27, 1973, the essential terms of the settlement remained the same. Once Nixon made the decision on October 6, 1972, to stick with the "decent interval" terms he and Kissinger had been pursuing for a year and a half, nothing he said about them thereafter was going to un-doom Saigon.

The Nixon Tapes

When I bought the best seller *The Nixon Tapes: 1971–1972* by Douglas Brinkley and Luke Nichter, the first transcript I looked up was from October 6, 1972. *The Nixon Tapes* leaves a blank space where the most important passages belong:

> Kissinger: And Thieu says that, sure, these proposals keep him going. Somewhere down the road he'll have no chance except to commit suicide.
>
> Kissinger: I've had a study.[1]

The Miller Center transcribed this tape as well. I've italicized the words that Nichter and Brinkley left out.

> Kissinger: And Thieu says that, sure, this—these proposals keep him going, but somewhere down the road he'll have no choice except to commit suicide. *And he's probably right. I mean, we—*
> President Nixon: *Let's talk among ourselves [unclear].*
> Kissinger: *We have to be honest—*
> President Nixon: *Right.*
> Kissinger: *—among ourselves.*[2]

Twice in the same conversation, the national security adviser told the president that their settlement terms would destroy South Vietnam. The second instance doesn't appear in *The Nixon Tapes*, either.

> Kissinger: *But—and I also think that Thieu is right, that our terms will eventually destroy him.*[3]

As for the president's response that if Saigon could not survive, "the thing to do is to look as well as we can"—it, too, is missing from *The Nixon Tapes*.[4] The book withholds from its readers the relevant evidence of Nixon's "decent interval" exit strategy. It dismisses the topic in a single, inaccurate sentence: "Yet some scholars argue that Nixon and Kissinger's strategy in Vietnam was never more than securing a 'decent interval' between American withdrawal and Communist takeover, a point that other scholars dispute, as have Nixon and Kissinger."[5]

After studying the subject for more than a decade, I don't know of *any* "scholars [who] argue that Nixon and Kissinger's strategy in Vietnam was *never more than* securing a 'decent interval'" (italics added). *The Nixon Tapes* doesn't name one. The "decent interval" was, after all, only part of Nixon and Kissinger's strategy. Besides, Jeffrey Kimball, the leading scholar of the "decent interval," has written that Nixon and Kissinger adopted the "decent interval" strategy only about halfway through Nixon's first term, so he doesn't claim that they *never* sought more.[6] Instead of providing the relevant evidence for readers to decide for themselves whether Nixon and Kissinger negotiated a "decent interval," *The Nixon Tapes* gives them a straw man.

Cutting off a transcript right before Kissinger admitted—twice—that Nixon's settlement terms would destroy South Vietnam is just the most glaring example of evidence suppression in *The Nixon Tapes*. The August 3, 1972, conversation ("South Vietnam probably can never even survive") is absent entirely. Parts of the May 29, 1971, tape appear in *The Nixon Tapes,* but the Vietnam passages ("the only problem is to prevent the collapse in '72") disappear into another blank space.[7] The book omits all the conversations about timing American military withdrawal from Vietnam to the presidential election from February, March, and the first half of April 1971.

The Nixon Tapes does include the February 18, 1971, conversation that provides the first inkling on tape of the "decent interval" strategy. It distorts, however, the key passage. Here's how it appears in *The Nixon Tapes*:

> Kissinger: I ask for a meeting with Le Duc Tho. Then have it October 15, and tell him, "Look, we're willing to give you a fixed deadline of total withdrawal next year for the release of all prisoners and a cease-fire." What we can then tell the South Vietnamese, "You've had a year without war to build up."[8]

The attentive reader will notice that the last sentence makes no sense. Since Kissinger was speaking in February 1971, when the Vietnam War was still raging, it would be odd for him to plan to say in October 1971 that South Vietnam had just had a year without war. Kissinger was describing the proposal he would make to settle the war; no one settles a war that has been over for a year.

What Kissinger really said makes much more sense. Here's the Miller Center transcript:

Kissinger: I ask for a meeting with Le Duc Tho. Then have it October 15th and tell him, "Look, we're willing to give you a fixed deadline of total withdrawal next year for the release of all prisoners and a ceasefire." What we can then tell the South Vietnamese—they've got a year without war to build up.[9]

The national security adviser was acknowledging that their settlement would not end the war but would merely provide an interval of a year between the total withdrawal of American troops and the war's resumption—an interval during which Saigon could build up its forces. *The Nixon Tapes*'s distortion of what Kissinger said, like its omissions of other conversations, comes once again at the expense of the evidence of the "decent interval" exit strategy.

Nichter claims that *The Nixon Tapes* provides "a more honest portrayal" of its subject: "The thing about Nixon is that at some moments he's just as bad as his worst critics say. But he is also just as good at moments as his proponents say. Some will say that our book leads to a much more sympathetic view of what it's like to be president. It may seem more sympathetic. But it's a more honest portrayal."[10]

As I noted earlier, suppression of evidence can be deliberate or inadvertent. Either way, *The Nixon Tapes* fails to give its readers an honest portrayal of Nixon on the tapes because it suppresses the evidence of Nixon at his worst—at the times he put politics above the lives of American soldiers.

Interpretive Inertia

The concept of a "decent interval" barely appeared in public debate during the Nixon years.[1] It wasn't until 1977 that Frank Snepp popularized it by publishing *Decent Interval: An Insider's Account of Saigon's Indecent End Told by the CIA's Chief Strategy Analyst in Vietnam*. Snepp's subtitle makes his insider status clear, but even a high-level CIA analyst in Saigon lacked access to the classified documents (and tapes) that captured

Nixon and Kissinger's real exit strategy. His book was titled with both intentional and unintentional irony: *Decent Interval* argues that Nixon and Kissinger did *not* settle for a "decent interval":

> As the Vietnam cease-fire gave way to renewed warfare, some of Kissinger's critics charged that he had never meant for the agreement to work anyway, but was merely trying through its convolutions and vagaries to assure a "decent interval" between the American withdrawal and a final fight to the death between the two Vietnamese sides. This judgment, however, hardly did justice to Kissinger himself. While he may not have put together a truly workable peace, he was most certainly concerned about preserving a non-Communist government in South Vietnam and the semblance of accommodation between Saigon and Hanoi.[2]

The "semblance" part is right. Regarding the situation in Vietnam, Snepp had detailed, firsthand, expert knowledge; regarding the "decent interval" negotiations, he remained in the dark, where Nixon and Kissinger kept the rest of America.

In their respective best-selling memoirs, Nixon in 1978 and Kissinger in 1979 took advantage of their exclusive access to the classified documents (and, in Nixon's case, tapes) to prop up the Dolchstoßlegende. Both suppressed the evidence of the "decent interval." Kissinger denied it outright:

> Nor is it correct that all we sought was a "decent interval" before a final collapse of Saigon. All of us who negotiated the agreement of October 12 were convinced that we had vindicated the anguish of a decade not by a "decent interval" but by a decent settlement. We thought with reason that Saigon, generously armed and supported by the United States, would be able to deal with moderate violations of the agreements; that the United States would stand by to enforce the agreement and punish major violations; that Hanoi might be tempted also by economic aid into choosing reconstruction of the North if conquest of the South was kept out of reach; that we could use our relations with Moscow and Peking, in addition, to encourage Hanoi's restraint; and that with our aid the South Vietnamese government would grow in security and prosperity over the time bought by the agreement, and compete effectively in a political struggle in which without question it had the loyalty of most of the population. Perhaps the Vietnamese parties could even work out a peaceful modus vivendi.

We could not know that soon Watergate would nullify most of these assumptions. In blissful ignorance of the future we landed in Washington, near joyous that we had brought home both peace and dignity.[3]

The words "decent" and "interval" do not even appear next to each other in Nixon's memoirs. Later, in 1985's *No More Vietnams,* he denounced the whole idea as immoral:

> Other hawks suggested a different approach. They conceded to the doves that we should not have gone into Vietnam in the first place, but contended that now that we were there, we had no choice but to see it through. Our goal, they argued, should not be to defeat the enemy but to stay long enough so that after we withdrew there would be a "decent interval" before South Vietnam fell to the Communists. I believed that this was the most immoral option of all. As president, I could not ask any young American to risk his life for an unjust or unwinnable cause.[4]

The "immoral" part is right. For the two decades between Nixon's resignation and death, history suffered from lack of access to his tapes and White House documents. Until Nixon drew his last breath on April 22, 1994, his post-Watergate campaign to suppress additional evidence of his wrongdoing, foreign and domestic, proved all too successful. Afterward, the National Archives finally could start releasing it to the public. Before Jeffrey Kimball published *Nixon's Vietnam War* in 1999, authors just didn't give a "decent interval" much time. William Bundy, for example, considered deceit to be so central to Nixon's foreign policy making that he named his 1998 book on the subject *A Tangled Web.*[5] Yet Bundy's critical study mentions the "decent interval" the exact same number of times as Nixon's self-serving memoirs: zero.

After Kimball, things changed. University historians began to accept his evidence. They didn't, however, always think its implications through. Jeremi Suri's *Henry Kissinger and the American Century* noted that Kissinger both negotiated a "decent interval" and realized that Saigon would fall without American troops. Despite the new evidence, however, Suri stuck with an old interpretation of the impact of Nixon and Kissinger's strategy:

> [Kissinger] hoped that through secret talks he could mix promises, threats and charm to obtain an honorable settlement to the war. He was also prepared to force an agreement on America's South Vietnamese allies, who opposed any reduction in support for what

Kissinger clearly recognized was an unsustainable regime. Though he would later deny it, the NSC adviser received consistent information that Saigon could not sustain an American troop withdrawal. Kissinger hoped to use his secret talks with North Vietnam to forestall the collapse of the US-supported regime as American troops withdrew. By 1971 he and Nixon would accept a "decent interval" between U.S. disengagement and a North Vietnamese takeover in the south. Secret talks with Hanoi would allow Kissinger to manage this process, preserving the image of American strength and credibility.[6]

"Preserving the image of American strength and credibility" with whom, exactly? Not with those Nixon called "our enemies," the Communist rulers of Moscow and Beijing. To them, Kissinger provided secret assurances that the North could conquer the South without fear of Nixon intervening, as long as it waited a year or two after he withdrew American troops. Our Cold War adversaries received a clear signal that Nixon was willing to abandon South Vietnam, provided that he could shift blame for the consequences off of his own shoulders. He forfeited America's geopolitical credibility abroad to maintain his political credibility at home. In their furtive negotiations for a "decent interval," Nixon and Kissinger revealed themselves to the Communists as craven and treacherous in their relationship with a supposed ally. They showed that they could accept the reality of defeat as long as they could avoid the appearance of it in the eyes of American voters. This had to have an effect on Moscow and Beijing, one Suri just did not reckon with: "The negotiations between Kissinger and Le Duc Tho coincided with increased Soviet and Cuban interventions throughout Africa. Leaders in Moscow, in particular, perceived a weakening in U.S. power and resolve, despite Kissinger's efforts."[7]

It was *because of*, not despite, Nixon and Kissinger's efforts that the Soviets judged that the Americans would not deploy American power to defend Saigon once a "decent interval" had passed. Moscow didn't merely "perceive" this weakening of resolve; Nixon and Kissinger made it explicit in the "decent interval" negotiations. Nixon and Kissinger got the North to sign the Paris Accords in the first place by letting it know that it could conquer the South militarily as long as it waited an extra year or two.

This approach—naked before one's enemies, disguised before one's nation—discredits even the private statements Nixon and Kissinger made to each other about preserving American credibility. On August 3, 1972,

Nixon said that it would worry the Chinese if Saigon fell within a few months after he withdrew the last American troops.[8] But as we've seen, Zhou Enlai accepted the "decent interval" slowly, reluctantly, and only after repeatedly urging Nixon and Kissinger to get out of Vietnam faster. There's no reason to think that Zhou preferred to delay Communist victory in Vietnam for a couple of years. The "decent interval" was something he accepted as the price of American withdrawal and subsequent nonintervention.[9] Likewise, when Nixon said on October 6, 1972, that it was important that "our enemies, the neutrals, and our allies after we finish . . . are convinced that the United States went the extra mile in standing by its friends," he and Kissinger had already let "our enemies" in Moscow and Beijing know how long they had to wait before he would *not* stand by South Vietnam.[10] These were foreign policy rationalizations of decisions made for domestic political purposes.

Suri presented no evidence that Nixon and Kissinger preserved American credibility with allied, neutral, or Communist countries. Who believed South Vietnam would survive their settlement? Who believed that adding four years to the war had helped rather than hurt America in the world? Who considered America the winner of this war? Anyone who wants to claim that Nixon and Kissinger preserved "the image of American strength and credibility" needs to come up with evidence that foreign governments were actually fooled.

Also deserving of scholarly attention is Nixon and Kissinger's decision to blame Congress—publicly and vehemently—for the consequences of their own actions. Congress, after all, is part of the United States government, so saddling it with the responsibility for losing a war was yet another way they forfeited American credibility in the world. In this they were obviously self-serving, blaming Congress as an alternative to blaming themselves. They were just as obviously *not* serving their nation. What good could it do America in the world to portray Congress as poised "to snatch defeat from the jaws of victory," as Nixon did in *No More Vietnams*?[11] Any nation that believed him would have to perceive America as an undependable, even treacherous, ally. Once again, they burnished their image in the eyes of America at the expense of America's image in the eyes of the world.

Likewise, Suri failed to assess whether it was possible to both negotiate a "decent interval" and "obtain an honorable settlement to the war." Since a "decent interval" merely interrupts hostilities without actually ending them, can it accurately be called a "settlement to the war"? In what sense can sending Americans to die in an unwinnable war and secretly betraying an ally be called "honorable"? Even Nixon acknowledged, in the abstract, that a "decent interval" was immoral.

Henry Kissinger and the American Century drew praise from Nixon critics and defenders alike. Niall Ferguson could not be friendlier to Kissinger, who has given the conservative historian access to his White House diaries and letters to write his official biography. Ferguson declared Suri's work "surely the best book yet published about Henry Kissinger."[12] But Ferguson and all future historians who write about Nixon and Kissinger's foreign policy need to do what Suri didn't—grapple with the damage done by timing military withdrawal from the war to the president's reelection campaign, negotiating a "decent interval" to delay, not deny, Communist victory over an ally, and blaming the consequences on Congress.[13]

Last Days in Vietnam

So entrenched is Nixon's Dolchstoßlegende that people add embellishments to it unnoticed and (sometimes) unknowingly. One such embellishment arrived in movie theaters during the writing of this book. It came from the unlikeliest of sources: a member of the Kennedy family. During the Nixon administration, the only liberals of national stature who came close to getting Nixon right were related by blood or marriage to JFK. The documentary Last Days in Vietnam was written and produced by Rory Kennedy, the daughter of Robert Kennedy and the niece of John Kennedy, Ted Kennedy, and Sargent Shriver. Last Days in Vietnam, as its title suggests, focuses on the fall of Saigon in April 1975. It recycles some old fallacies about the war's final days before adding some new ones to the pile. Henry Kissinger gets plenty of screen time.

> Kissinger: We who made the agreement thought it would be the beginning, not of peace in the American sense, but the beginning of a period of coexistence which might evolve, as it did in Korea, into two states. Reconciliation between North and South, we knew, would be extremely difficult, but I was hopeful.[1]

Last Days would have served its viewers better by juxtaposing Kissinger talking about reconciliation on camera with his accurate predictions of postsettlement warfare and North Vietnamese military victory on the Nixon tapes. But the documentary doesn't include any Nixon tapes. Nor does it include an interview with anyone explaining that Nixon and Kissinger got the North to sign the Paris Accords by secretly allowing it to take over South Vietnam following a "decent interval." That would have enabled viewers to make sense of the North's March 1975 offensive, which took place a "decent interval" of two years after the last American soldiers left in March 1973.

Last Days misleadingly depicts that March 1975 offensive with an animated map. At first, the map shows no red (North Vietnamese) in South Vietnam. Then it shows red flooding into the South across the DMZ.

At another point viewers hear Frank Snepp estimate that 160,000 North Vietnamese troops threatened the Saigon government, but no one mentions that more than 90 percent of them—150,000—had been in the South since the ceasefire-in-place began two years earlier. If Snepp is right (and he did work in Saigon as a CIA analyst on estimates of enemy strength), then the total number of North Vietnamese troops in the South increased by less than 10 percent from the settlement's signing to Saigon's fall. The animation in *Last Days* makes it look like there was a 100 percent increase in one month. It heightens the drama, but distorts the history.

These omissions just exaggerate the existing Dolchstoßlegende; another set of omissions from *Last Days in Vietnam* adds a whole new fictional chapter to it.

Last Days cuts out essential information about the struggle between President Ford and Congress over a last-minute aid request. In the excerpt below from the film's portrayal of Ford's April 10, 1975, address to Congress, I've indicated where the filmmaker omitted portions of the address:

> President Ford: The situation in South Vietnam [*text omitted*] has reached a critical phase requiring immediate and positive decisions by this government. [*text omitted*] There are [*text omitted*] tens of thousands of South Vietnamese employees of the United States government, of news agencies, of contractors and businesses for many years whose lives, with their dependents, are in very grave peril. [*text omitted*] I am therefore asking the Congress to appropriate without delay $722 million for emergency military assistance [*text omitted*] for South Vietnam. [*text omitted*] If the very worst were to happen, at least allow the orderly evacuation of Americans and endangered South Vietnamese to places of safety.[2]

The casual viewer could be forgiven for assuming that the $722 million emergency military aid request was needed to evacuate Americans and endangered South Vietnamese. But that was not so.

For this reason, it is misleading for *Last Days* to show White House Press Secretary Ron Nessen saying:

> Congress wouldn't pass it. They said, "No more." You know? "No more troops, no more money, no more aid to the Vietnamese." Well, I had to go in to President Ford's office to tell him. I had never heard Ford use a curse word in all the time I'd known him. But when I showed him this story, he said, "Those sons of bitches."[3]

Again, the casual viewer of *Last Days* could be forgiven for assuming that Congress did not vote for aid to evacuate Americans and endangered South Vietnamese. Again, this just is not so.

In reality, Ford made *two* aid requests. Here's the unedited quote from Ford's April 10, 1975, address to Congress with the words that *Last Days* left out set in italics:

> I am therefore asking the Congress to appropriate without delay $722 million for emergency military assistance *and an initial sum of $250 million for economic and humanitarian aid* for South Vietnam.[4]

The humanitarian aid covered the costs of evacuating Americans and endangered South Vietnamese, and by April 24 both houses of Congress had approved it.[5] Within twenty-four hours of doing so, House and Senate negotiators reached a compromise—on a *higher* level of humanitarian aid than Ford had requested.

> CONFEREES AGREE ON $327 MILLION EVACUATION BILL Washington (UPI)—Senate and House conferees reached formal agreement today on legislation authorizing $327 million to finance humanitarian relief and the evacuation of Americans and friendly Vietnamese from Saigon....
>
> At the White House, Mr. Ford was described as "pleased" by the compromise legislation.
>
> Press Secretary Ron Nessen said Mr. Ford had notified congressional leaders that he was taking advantage of a provision in the 1975 Foreign Assistance Appropriations Act allowing him to temporarily transfer money from an "Indochina postwar reconstruction fund" for use in meeting the evacuation cost.[6]

Last Days in Vietnam makes *no* mention of the humanitarian aid request or of the congressional votes to use it to evacuate Americans and endangered South Vietnamese. Instead, the documentary shows Snepp saying, "Young officers in the embassy began to mobilize a black operation, meaning a makeshift underground railway evacuation using outgoing cargo aircraft that would be totally below the radar of the ambassador." That may have been necessary, especially before President Ford requested—and the House and Senate okayed—authorization to evacuate endangered South Vietnamese. But a full-fledged, official, *overground* operation was soon under way.

> U.S. SEEKS TO SPEED FLOW OF REFUGEES FROM SAIGON | GOAL PUT AT 8,000 A DAY Washington, April 24—As fears rose here that time was running out on plans for an orderly evacuation of Americans and Vietnamese from Saigon, the director of President Ford's Refugee Task Force said today that he was trying to accelerate the outflow.
>
> Ambassador L. Dean Brown, who is coordinating the inter-agency relief effort, said in a news briefing at the State Department that about 5,000 persons were being flown daily to Guam, most of them in American military aircraft, but that he hoped "to see that raised enormously."
>
> "I'd like to see them up around 8,000 to 9,000," he said, adding that if Communist forces started shelling Tan Son Nhut air base near Saigon, the airlift would have to cease.
>
> "That could happen at any moment," he said, but he declined to speculate on how much time remained to complete the evacuation of as many as 130,000 South Vietnamese, 50,000 of whom would be admitted to the United States on "parole" status as "high risks" whose lives might be endangered if they remained in Vietnam.[7]

Far from being a secret black op, the evacuation in Saigon's final days was overt, authorized by the president and Congress, and reported on the front page of the *New York Times*. Once again, *Last Days* heightens the drama at the expense of the truth.

It's true that Congress rejected Ford's separate, $722 million request for military aid. To its credit, *Last Days* has Kissinger acknowledge how little impact that aid would have had.

> Kissinger: We knew we were not going to get the $722 million. By that time, it made no big difference. But President Ford said he owed it to Vietnam to make a request.[8]

While the $722 million request would not have saved Saigon, it did furnish Kissinger, Nixon, and Ford with one more excuse to shift blame for its fall onto Congress. "I am absolutely convinced," President Ford declared on April 16, 1975, "if Congress made available $722 million in military assistance in a timely way by the date that I have suggested—or sometime shortly thereafter—the South Vietnamese could stabilize the military situation in South Vietnam today."[9] It had taken American advisers on the ground coordinating air power to stabilize the military situation during the 1972 Easter Offensive. President Ford didn't propose sending any American forces back into battle—just money, equipment, and supplies. As Murrey Marder reported in the *Washington Post* on April 11, 1975: "There is little or no real prospect, in the judgment of administration strategists, that the requested $722 million in new military aid for South Vietnam could redress the military balance. Indeed, it is unlikely that this aid could reach South Vietnam in time to avert disaster for the government."[10]

My own view is that Congress should have responded to Nixon and Ford's aid requests for Saigon by doubling them. The money would have been wasted in South Vietnam, but it would have been a small price to pay—the per-taxpayer equivalent of pocket change—to demonstrate that Saigon was not going to survive no matter how much money America threw at it. The price of living with the backstabbing myth has been much, much higher. Americans at the time, however, were understandably disinclined to throw their money away. "A lopsided 81-to-12 percent," the pollster Louis Harris reported at the time, were "opposed to President Ford's request for $722 million to subsidize military aid to Saigon"—a majority large enough to include the Left, the Center, and even a majority of the Right.[11]

The omissions of *Last Days* left viewers and reviewers seriously confused. "When the South indeed became vulnerable," the movie reviewer Ann Hornaday wrote in the *Washington Post,* "a war-weary Congress and the American populace they represented effectively put the kibosh on more aid, either for military support or an orderly evacuation."[12] False—Congress passed more evacuation aid than Ford requested; that wasn't the problem. "Ambassador Graham Martin, a rigid Cold Warrior out of 'The Quiet American,'" wrote George Packer for the *New Yorker,* "refused to believe that Saigon was about to fall, and wouldn't allow fixed-wing air evacuations from the Tan Son Nhut airbase while it remained out of North Vietnamese hands."[13] False—Martin allowed thousands of South Vietnamese to fly out on fixed-wing aircraft every day until North Vietnamese attacks on Tan Son Nhut airbase eliminated that option; that wasn't the real problem, either.[14]

The real problem was that the White House didn't start the evacuation until mere weeks before Saigon fell. The time to make provisions for the safe passage out of the country for endangered South Vietnamese was during the same settlement negotiations that provided safe passage out for the last American troops. That, after all, was when the fall of Saigon became a foregone conclusion. But Nixon and Kissinger left the South Vietnamese to fight it out with the North and lose, because to do otherwise would have required an admission that the president and his foreign policy alter ego had never come up with a way to win the war. They lost it in those negotiations; they should have taken responsibility then for the safety of those who had fought on America's side or served it in a civilian capacity. But Nixon and Kissinger were willing only to accept defeat, not to admit it. Once again, they put their political interests above the lives and safety of the people depending on them.

Last Days in Vietnam also suffers from a curious lack of on-screen Kennedys. It would have done more to clarify why Saigon fell if it had shown Ted Kennedy in 1971 saying, "The only possible excuse for continuing the discredited policy of Vietnamizing the war, now and in the months ahead, seems to be the president's intention to play his last great card for peace at a time closer to November 1972, when the chances will be greater that the action will benefit the coming presidential election campaign."[15] A clip of Sargent Shriver's assessment of the draft settlement—"I don't see what's the difference between what he has got and what he used to call surrender"—would also have helped. A discussion of John, Robert, and Ted Kennedy's positions on a coalition government might have illuminated an alternative way to end the war—one that could have included emigration provisions to give friendly South Vietnamese an alternative to dying in a bloodbath or living in a reeducation camp.

In contrast, it could have shown the Republican nominee publicly rejecting a coalition as "disguised surrender" in 1968, then negotiating a better-disguised surrender as president.[16] There's no way to understand America's last days in Vietnam without seeing how the deal Nixon and Kissinger made with the Communists set Saigon up to fall.

A Better War

The prevailing popular interpreter of Nixon's policies is Lewis Sorley, author of *A Better War: The Unexamined Victories and Final Tragedy of America's Last Years in Vietnam*. Sorley can make Nixon's case without Nixon's baggage. *A Better War* tiptoes right to the edge and stops just short of saying that Nixon led America to victory in Vietnam: "There came a time when the war was won. The fighting wasn't over, but the war was won."[1] It's an oddly passive way to declare victory, but Sorley has a bad habit of skirting the subjects of his own sentences:

> Not only was the internal war against subversion and the guerrilla threat won, so was that against the external conventional threat—in the terms specified by the United States. Those terms were that South

Vietnam should, without help from U.S. ground forces, be capable of resisting aggression so long as America continued to provide logistical and financial support, and—of crucial importance later, once a cease-fire agreement had been negotiated—renewed application of U.S. air and naval power should North Vietnam violate the terms of that agreement.[2]

When Sorley writes of "the terms specified by the United States," he doesn't identify either the man who did the specifying or the essential context. The terms were Nixon's. Publicly, Nixon claimed that Saigon could survive without American ground troops, provided it had American aid and airpower (and offshore naval bombardment of the North). Privately, he didn't count on it—not when he decided to keep American troops in Vietnam until his reelection was safe, not when he had Kissinger negotiate a "decent interval," and not when he offered to let Congress tie his hands. Nixon's real terms were quite different from his public ones.

Referring to Nixon as "the United States" allows Sorley to slide past one of the problems with his thesis: the subject of his book, Gen. Creighton Abrams, was on the record saying that South Vietnam could *not* survive without American ground troops. So were the CIA, Pentagon, Joint Chiefs, and all the other participants in NSSM-1. Given the consensus of American officialdom on this point, for Sorley to refer to Nixon's public terms as "the terms specified by the United States" distorts the views of the Nixon administration's top officials—including the topmost two, Nixon and Kissinger. Sorley doesn't even mention NSSM-1, although he does acknowledge in passing that the Military Assistance Command, Vietnam (MACV, the command Abrams led) had reached the "conclusion in the autumn of 1969 that 'unless North Vietnamese forces return to North Vietnam, there is little chance that any improvement in RVNAF [the Republic of South Vietnam's Armed Forces] or any degree of progress in pacification, no matter how significant, could justify significant reductions in U.S. forces from their present level.'"

"In other words," Sorley writes, "their assessment was that, despite the gains being made by South Vietnam's forces, their situation would become untenable if the NVA [North Vietnamese Army] were permitted to remain in the South while U.S. forces withdrew. That conclusion was, of course, going to vanish in the wind of domestic pressures for withdrawal."[3] Facts vanish in the wind of Sorley's metaphor.

Once more, passive phrasing ("if the NVA were permitted to remain in the South") provides Sorley with a verbal escape route from a big problem with both his thesis and the war itself: Abrams never came up with a way to drive and keep the North Vietnamese out of the South. Neither did Nixon or any of his military and civilian advisers. This is another way of saying that no one ever came up with a way to win the war. Sorley stated the problem well in his earlier book, *Thunderbolt: General Creighton Abrams and the Army of His Times*:

> [Saigon] was asked to sign on to an agreement that would sanction the continuing presence in the South of hundreds of thousands of invading troops. Agonizing as it was, their ouster had proven too difficult to attain, no less so in Paris than on the battlefield.
>
> Thieu sensed—correctly—that a ceasefire in place foreshadowed the eventual downfall of an independent South Vietnam. Henry Kissinger maintained—perhaps also correctly—that he got the best agreement possible under the circumstances. In *that gap between what was necessary and what was attainable* lay the essential tragedy of the war in Vietnam.[4]

It's hard to argue with the logic: If (1) South Vietnam needed American troops backing it up in order to withstand attack by North Vietnamese troops, and (2) no one found a military or diplomatic way to get and keep North Vietnamese troops out of the South, then (3) the war was unwinnable. Nixon's choice was either to keep American ground forces fighting and dying in Saigon's defense, or to leave and lose. How odd that the man who became famous for the argument that the war was won came up with the simplest, clearest way to state that the war was unwinnable.

How much odder that fans of *A Better War* didn't notice. Sorley didn't abandon his earlier conclusion so much as downplay it to the point of near-invisibility:

> The key element in the outcome of the war was the continued presence of North Vietnamese armed forces in South Vietnam. The Paris Agreement was silent on that point, thus allowing the NVA to maintain in place the forces they had always denied were in South Vietnam. This silence constituted a fatal flaw, at least from the South Vietnamese perspective, *even if it was also an inevitable acceptance of the battlefield realities.* "That ceasefire agreement," wrote Sir Robert Thompson, "restored complete security to the [enemy] rear bases in North Vietnam, in Laos, in Cambodia, and in the parts of South Vietnam that it held. It subjected the South Vietnamese rear base again to being absolutely open to military attack. That is what the cease-fire agreement actually achieved."[5]

A Better War brought great public attention to the shift in American strategy from "search and destroy" under Gen. William Westmoreland to "clear and hold" under Gen. Creighton Abrams. To put it simply, "clear and hold" is the name for tactics aimed at securing a population by clearing enemy troops out of an area and holding it with American troops (to be replaced by local forces) so the enemy cannot return. In Vietnam, "clear and hold" tactics produced results attested to by many Americans who remarked at their ability to travel safely through parts of South Vietnam that previously had been dangerous. To the unwary, "clear and hold" might sound exactly like a solution to the problem of how to drive and keep the North Vietnamese Army out of the South. Compare the situation on the ground before and after Abrams implemented "clear and hold." In 1968, the year Abrams took command, American intelligence estimated the total number of soldiers the North Vietnamese had sent into the South since the first of the year at 150,000.[6] In December 1972, the month before the signing of the Paris Accords, after four years of "clear and hold" tactics, American intelligence estimated the total number of North Vietnamese soldiers in the South at 150,000.[7] "Clear and hold" may have been a better tactic, but it was not a strategy that could win the war.

If Sorley's readers had only realized that Nixon- and Abrams-era tactics had merely lost Vietnam less conspicuously, they might not have been so eager to adopt the same tactics for Iraq in 2007 and Afghanistan in 2009.[8]

The Aid-Cutoff Myth

During the 2005–6 debate over whether to withdraw from Iraq, Nixon's former defense secretary concocted another false lesson of history: "During Richard Nixon's first term, when I served as secretary of defense, we withdrew most U.S. forces from Vietnam while building up the South's ability to defend itself. The result was a success—until Congress snatched defeat from the jaws of victory by cutting off funding for our ally in 1975."[1]

Despite what Melvin Laird wrote in 2005 for *Foreign Affairs,* the prestigious journal of the Council on Foreign Relations, Congress never did cut off aid to South Vietnam. Even Nixon and Sorley acknowledged that it appropriated $700 million in aid to Saigon for fiscal year 1975.[2] This was less than half of the $1.47 billion that Nixon had requested, but it wasn't nothing.

Telling such a big whopper in such a public forum might seem like a good way to get caught, but Laird didn't. On January 17, 2007, he repeated the falsehood in an even more prominent forum: the *Washington Post* op-ed page:

> The brewing fight in Congress over continued funding of the war in Iraq will not be the country's first. It is an ominous reminder of 1975, when Congress cut off funding for the Vietnam War three years after our combat troops had left. With the assistance we promised South Vietnam in the 1972 Paris Accords—U.S. equipment, replacement parts and ammunition—it had won every major battle since we left. But Congress lost the will to keep our promise and killed the appropriation. The result was a bloodbath.

There was a bloodbath, but it didn't result from an imaginary congressional aid cutoff, or even from the genuine cuts Congress made in Nixon's aid request. It resulted from Nixon's "decent interval" exit strategy.

Laird demonstrated the power of political myth to warp policy debates. In *Foreign Affairs,* he urged policy makers to adopt an "Iraqization" program of training and equipping Baghdad's army.[3] He portrayed Vietnamization as a victorious strategy to be used as a model. In the *Post,* he urged the new, Democratic Congress not to do what many of its members had been elected to do: bring American troops home from Iraq. (Laird characterized withdrawing the troops as cutting off funding for the war, which is misleading. As we have seen, long after American troops came home from Vietnam, America continued to fund the war waged by South Vietnam's military. Withdrawing and cutting off funds are two different things.)

Armed with Nixon's Dolchstoßlegende, Laird and his allies won their policy battles in Washington. Sadly but predictably, their policies once again failed to win a war. In 2007, President Bush vetoed troop withdrawal legislation and instead launched "the surge," a sharp, temporary increase in American troops in Iraq. In 2008, he negotiated an agreement with the Iraqi government to allow American troops to remain another three years past the end of his presidency. These five additional years of war—five years of American training and equipping the Iraqi army—did not stop four divisions of the Iraqi army from dropping their weapons, tearing off their uniforms, and fleeing before the irregular insurgents of the Islamic State in June of 2014.[4] As in Vietnam, the American-trained forces in Iraq greatly outnumbered the foes from whom they fled. They were well-equipped, but they let their equipment fall into the hands of their attackers.[5] "They are crumbling," said James Dubik, the lieutenant general in charge of training the Iraqis during the surge.[6]

The crucial difference between *training* people to fight and *winning their loyalty* to our side of a fight revealed itself in Iraq, just as it had decades earlier in Vietnam. It was one more lesson of Vietnam obscured by Nixonian myth.

How Wars Don't End

Failure to see through Nixon's Dolchstoßlegende can be fatal. It leaves Americans—particularly the ones in uniform—vulnerable to politicians who would do as Nixon did. One pillar of the foreign policy establishment publicly advised President Barack Obama to use Nixon as a role model in Afghanistan:

> What he needs is a strategy for getting out without turning a retreat into a rout—and he would be wise to borrow one from the last American administration to extricate itself from a thankless, seemingly endless counterinsurgency in a remote and strategically marginal region. Mr. Obama should ask himself, in short: What would Nixon do? Richard M. Nixon and his national security adviser, Henry A. Kissinger, tried to manage the risks of exiting the Vietnam War by masking their withdrawal with deliberate deception and aggressive covering fire.[1]

The author—Gideon Rose, editor of *Foreign Affairs*—left unspecified the exact nature of the "deliberate deception" perpetrated by Nixon and Kissinger. "The Nixonian approach," Rose wrote, "has its costs: it would generate charges of lying, escalation and betrayal." That it should do, since the Nixonian approach involved lying (the false claim that Vietnamization would make Saigon capable of defending itself, the related false claim that Nixon was not timing American military withdrawal to the 1972 election, and other false claims detailed in these pages); escalation (of the politically popular, but strategically ineffective, bombing-and-mining sort); and betrayal (of all the American soldiers Nixon kept in Vietnam to secure his reelection, of their families, and of American voters as a whole). The real problem with the Nixonian approach is not that "it would generate charges of lying, escalation and betrayal," but that the charges would be true. Rose counted only the costs that would be borne by a politician who chose to take his advice; the real costs of a Nixonian approach would be borne by the nation, particularly its armed forces. Rose neglected to mention all the lives lost so Nixon and Kissinger could delay Saigon's fall past Election Day 1972 and, via a secretly negotiated "decent interval," past Election Day 1974 as well. He also neglected to mention the loss of American credibility inherent in secretly assuring the Communists that if they waited a "decent interval" before taking over the South, Nixon would not intervene to save it. By not spelling out the nature of Nixon and Kissinger's "deliberate deception," Rose left his readers unable to judge its costs and consequences.

That didn't stop the *New York Times* from putting Rose's advice on the front page of its very first "Sunday Review" section. How did Rose manage to urge Obama to follow Nixon's example without ever quite saying what that example was? By quoting, unattributed and in truncated form, a transcript I made of the August 3, 1972, Oval Office tape in which Kissinger said that "we've got to find some formula that holds the thing together a year or two, after which . . . no one will give a damn." I intended the transcript to serve as an exposé, not a how-to.[2]

Rose claimed that the president continued and escalated the war "to buy time and space for its 'Vietnamization' programs to work." As we've seen, Nixon kept America in the war for four more years precisely because he realized Vietnamization would *not* work. Rose was certainly familiar with the evidence; in his book *How Wars End: Why We Always Fight the Last Battle,* Rose cited an article I wrote in *Diplomatic History* on the subject. My point was hard to miss:

> President Richard M. Nixon timed American military withdrawal from Vietnam to the 1972 U.S. presidential election. He kept American troops in Vietnam into the fourth year of his presidency to avoid a South Vietnamese collapse prior to Election Day 1972—a collapse that would have demonstrated that his "Vietnamization" program of training and equipping the South to defend itself had failed. By early 1971, he had decided to stretch the withdrawal out so that the last troops would not come home until shortly before or after the presidential election. His secret timetable served the political purpose of concealing Vietnamization's failure long enough to render voters incapable of holding him accountable for it. . . . Nixon was determined to stick to his election-centric timetable regardless of whether Hanoi agreed to a settlement (something Nixon thought he had only a 40 to 55 percent chance of getting) and regardless of whether South Vietnam could survive without American troops.[3]

For some unknown reason, Rose altered the title of my article (which is the same as the title of this book) to "Fake Politics." I must admit that Rose's inaccuracy has a certain poetry, given the article's subject matter. Unfortunately, it exemplifies Rose's tendency to distort and obscure key aspects of Nixon's strategy—particularly Nixon's decision to sacrifice the lives of American soldiers to delay Saigon's fall until after he was reelected. The "Nixonian approach" was not just fake; it was *fatal*. Most people expect a degree of fakery from politicians (who must smile on cue and say they're glad to be in Insert Name of Town Here, even when they're not), but most would reject as immoral the "Nixonian approach" of prolonging a war for political gain—at least when they finally got to hear about it. They didn't hear about it from Rose in the pages of the *Times*.

Rose didn't *deny* that Nixon timed American military withdrawal to his reelection, or that Nixon did this because he thought that, even after Vietnamization, Saigon would still need American troops to survive. (Since my article appeared in *Diplomatic History* in 2010, no scholar has published any attempt to refute its argument that Nixon secretly adopted an election-oriented withdrawal timetable.) Rose just didn't mention Nixon's secret timetable. The timetable is, however, relevant and compelling evidence that not even Nixon himself believed that Vietnamization would work.

In the *Times,* Rose fuzzed the issue of whether Vietnamization worked, along with other key questions:

> [A] rule of withdrawal is to remain engaged, providing enough support to beleaguered local partners so they can fend off collapse for as long as possible. Withdrawal should be defined as the removal of ground forces from direct combat, not the abandonment of the country in question.
>
> The Nixon administration tried to do this, and its success in stopping North Vietnam's Easter offensive in 1972 showed that it could work. But once American troops and prisoners came home, few displayed any appetite for reengagement. Congress ordered an end to all military operations in Southeast Asia and cut aid to Saigon, making its eventual collapse a foregone conclusion.

Rose subtly lowered the bar for Vietnamization from Nixon's public promise that it would enable Saigon to defend itself to his private acknowledgment that it would merely delay defeat—or as Rose put it, "fend off collapse for as long as possible." Rose also evaded the question of whether Saigon could survive without American ground troops, since he redefined withdrawal as "the removal of ground forces from direct combat." Technically, the American advisers who made it possible for South Vietnam to survive the Easter Offensive were not combat troops, but they were "ground forces." They were not, however, "removed from direct combat"—the North Vietnamese still shot at them. By evading the issue of whether Saigon needed American advisers on the ground to survive, Rose could also evade the related issue of Hanoi's refusal to release the POWs unless Nixon withdrew *all* American ground troops, advisers included. Because Nixon agreed to complete American withdrawal to gain the POWs' release, the South Vietnamese after 1973 would not have the American advisers who played a crucial role in stopping the 1972 Easter Offensive—yet another issue Rose evaded.

Rose's worst evasion, however, was blaming Congress for "making [Saigon's] eventual collapse a foregone conclusion." Nixon and Kissinger did that themselves.

Rose didn't deny that they negotiated a "decent interval." In fact, in *How Wars End,* he cleverly juxtaposes Kissinger's denial with contradictory evidence from the NSC transcripts of conversations with Zhou and Gromyko, Kissinger's Polo I briefing book marginalia, and the August 3, 1972, transcript. Unfortunately, Rose failed to think through the consequences of negotiating a "decent interval"—the most important (and obvious) one being that the North would conquer the South once the interval ended. Instead, Rose endorsed Nixon's Dolchstoßlegende: "Had events in Washington played out differently—with Watergate not crippling the administration and with Congress less hell-bent on slamming the door behind the departing U.S. ground troops—Nixon might have been able to send enough aid and bombs to keep the Thieu regime in power."

If it were only that simple. How could Nixon gain the release of the new POWs Hanoi would capture by shooting down American bombers (since he'd no longer be able to trade the withdrawal of American ground troops)? How could bombing the North after 1973 stop the flow of soldiers and supplies into the South when bombing before 1973 had failed to do so? In the absence of American advisers, who would get the South Vietnamese army to stand and fight long enough to force North Vietnamese soldiers to amass in great enough numbers to provide "lucrative targets" for American bombers? And without American advisers to call in the targets, how would American bombers know where to strike? Why *wouldn't* Saigon fall without American ground troops, as General Abrams, the CIA, the Pentagon, the State Department, and the Joint Chiefs predicted it would in NSSM-1? What about the risk that the Chinese or Soviets would play tapes of Kissinger assuring them that Nixon would *not* intervene? These are questions Rose failed to raise, much less answer. (Rose has both the practical and intellectual background to address such questions. He is former associate director for Near East and South Asian affairs in President Bill Clinton's NSC, and the former assistant editor for two leading publications of the neoconservative movement, *The Public Interest* and *The National Interest*.)

Rose accused others of intellectual dishonesty:

> As Senator John Kerry put it to Kissinger in 1992, "what you wound up with in 1973 was extraordinarily close to the program tabled in 1969." This argument comes in both a dishonest and an honest form. The dishonest one is to say, in effect, that 1973 could have been achieved in 1969. That is simply not true, since a key element of the 1973 settlement—Thieu's short-term and possibly long-term survival—was not available earlier. The honest form is to say that 1973 implied 1975, and that 1975 could (and should) have been had in 1969. This puts the real issue squarely on the table, and whether one accepts the argument depends on how important one thinks it was (for whatever reasons) for Washington to give South Vietnam a shot at lasting past American withdrawal.

Even while claiming that others were evading the real issue, Rose managed to evade the real issue. It wasn't whether "to give South Vietnam a shot at lasting past American withdrawal." South Vietnam would last past an American withdrawal—either for a year or two with a "decent interval" deal or for a few months without one. The real issue was whether delaying Saigon's death for a couple of years was worth the deaths of more than 20,000 Americans and many more thousands of Vietnamese. *That* is the honest form of the argument.

Rose fuzzes the issue by tossing in something vague about Thieu's "possibly long-term survival." That "possibly" is wiggly. Possibly, peace will break out in the Middle East tomorrow; no one, however, should count on it. Given the evidence that Nixon, Kissinger, and Thieu did *not* believe in South Vietnam's "long-term survival" in the absence of American troops, Rose needs to come up with evidence that they (along with the JCS, Pentagon, CIA, etc.) were all wrong. The truism that anything's possible is just not enough.

Questions Unasked

This brings us to the last and worst consequence of Nixon's Dolchstoßlegende. Americans are defenseless to stop another president from doing what Nixon did. Actions taken by both President George W. Bush and President Barack Obama raise the question of whether they merely postponed, rather than prevented, defeat in Iraq and Afghanistan.

Bush's 2007 surge in Iraq would have required a miracle to succeed. The premise was that temporarily increasing the American troop presence and employing "clear and hold" tactics to secure the population would provide diplomats an opportunity to work out a settlement among Iraqi parties who were then shooting out their differences. The miracle required was to turn blood enemies into peaceful, democratic partners. It did not occur, of course, nor did anyone come up with a workable way to make it occur. The surge did, however, solve the most dire political problem facing Bush and the Republican Party. Instead of withdrawing American troops and acknowledging failure, the surge enabled them to claim "progress" of the kind Nixon made in Vietnam. The increased American troops did increase the security of the Iraqi population—temporarily. But the surge troops started coming home before the Bush administration ended. As the number of American troops gradually decreased during the next three years, violence among the Iraqis increased once more.[1] Geopolitically, the surge produced temporary, tactical benefits, but it failed to achieve its strategic goal. Politically, however, it provided the Republican president with an excuse to continue the war long enough to hand it off to his successor. President Bush never did find a way to win the war, but he did find a way to shirk responsibility for losing it.

Nevertheless, House Speaker Nancy Pelosi, D-California, reached back into the McGovern playbook to accuse the president of "an open-ended commitment to a war without end."[2] Once again, a Democratic leader was casting a Republican president as someone who would never retreat, never surrender. On the other hand, Chairman Joe Biden, D-Conn., of the Senate Foreign Relations Committee, did accuse the president of merely delaying disaster so his successor would "be the guy landing helicopters inside the Green Zone, taking people off the roof." Like Ted Kennedy in 1971, Biden in 2007 lacked documents (or tapes) to back up his claim. "I have reached the tentative conclusion that a significant portion of this administration, maybe even including the vice president, believes Iraq is lost," said the future vice president. "Therefore, the best thing to do is keep it from totally collapsing on your watch and hand it off to the next guy—literally, not figuratively."[3]

Bush also handed his Democratic successor a particularly Nixonian withdrawal deadline: December 31, 2011, Reelection Year Eve for the next president. Recall that Nixon rejected a Reelection Year Eve withdrawal deadline of December 31, 1971, because it would have allowed Saigon to unravel before Election Day and voters to hold him accountable. The

December 31, 2011, deadline accomplished two ends helpful to Republicans: It kept the war going long enough for it to become a Democratic president's responsibility, and it provided almost a year of unraveling in Iraq before Election Day 2012.

"A Romney for President White Paper" from October 7, 2011, laid the groundwork for a new Dolchstoßlegende:

> The 2007 "surge" of troops successfully provided security to the population and granted space and time for the Iraqis, our diplomatic corps, and our coalition partners to establish institutions of governance. Today, after struggle and sacrifice, the goal of a democratic Iraq allied with the United States is within our reach. The Obama administration, however, is threatening to snatch defeat from the jaws of victory. We are nearing the December 31, 2011 deadline for reaching a new Status of Force Agreement that will allow U.S. troops to remain in Iraq to continue their training mission.[4]

Far from creating a democratic ally of the United States, the invasion of Iraq and subsequent efforts at nation building replaced the minority Sunni-dominated government of Saddam Hussein with a majority Shiite-dominated government friendly to Iran. At the time of Romney's white paper, the debate was over how many American troops to keep in Iraq past the deadline. Romney accused Obama of planning to keep too few there.[5] The Iraqi government resolved the debate by refusing to allow any American troops to remain in the country past the December 31, 2011, deadline.

Voters familiar with the true story of Nixon and Vietnam would have had other questions on their minds than how many troops to keep in Iraq: Was the president's proposal to keep American troops in Iraq past the December 31, 2011, deadline an attempt to keep the country from unraveling before Election Day 2012? Would keeping American troops in Iraq another five years achieve the goal that the previous five years had not? Was the opposition party accusing the president of doing too little too late because that's a time-tested way for the party out of power to become the party in power? Were any of the proposals to keep America militarily involved in Iraq, from the surge through the deadline-extension debate, part of a workable strategy to transform the country into a democratic ally of the United States? Or did they amount to little more than useful ways for politicians to delay, and shift the blame for, defeat?

President Obama adopted a distinctly Nixonian withdrawal date for the 33,000 surge troops he sent to Afghanistan: September 2012, two months before the election. He kept them in long enough to avoid unraveling before Election Day and brought them home in time to take credit for meeting his goals before voters went to the polls. Obama also set, years in advance, a timetable for withdrawing most American troops from Afghanistan by the end of 2014. That was enough to get him through the last congressional midterm election of his administration without losing the war.

More than half of the American casualties of the war in Afghanistan have occurred during the Obama administration.[6] No one can say that these sacrifices have led to victory in what is now America's longest war.

None of this proves that Bush or Obama made military decisions for political reasons. The point is that voters don't even know that there are questions that must be raised whenever a president proposes to prolong a war without a workable strategy to win. If the Nixon tapes are, in Bob Woodward's witty phrase, the gift that keeps on giving, then Nixon's Dolchstoßlegende is the gift that keeps on taking. The myth disguises political cowardice as political courage, opportunism as patriotism, and defeat as victory. It keeps us from the knowledge of our national past that we need to make wise decisions for the future. For as long as the true story of how Nixon prolonged a war and faked peace for political gain remains unknown, other presidents will be able to play fatal politics with American lives.

ACKNOWLEDGMENTS

Many people helped make this book possible. The University of Virginia's Miller Center under the visionary leadership of Gov. Gerald L. Baliles has provided me with crucial institutional support for years of research and writing on the White House tapes and the Vietnam War. The gifted scholars of the Presidential Recordings Program (PRP)—David Coleman, W. Taylor Fain, Patrick J. Garrity, Kent Germany, Max Holland, Erin R. Mahan, Ernest R. May, Guian A. McKee, Timothy J. Naftali, Marc Selverstone, Dave Shreve, and Philip D. Zelikow—have contributed enormously to the nation's and my own understanding of the priceless history contained in these secretly made recordings. PRP Chairman Marc Selverstone and Assistant Editor Keri Matthews did the vital, time-consuming, and unsung work of reviewing, revising, and perfecting the transcripts excerpted in this book under a tight schedule, and they did so with great skill, expertise, and patience. (The PRP interns, who toil anonymously on draft transcripts, are the scholarly stars of tomorrow, so I thank them here as well in the hope they will treat me kindly when they begin their reign.) Thank you also to Pat Dunn and Lorraine Settimo for their patience and wisdom. Many people at the Miller Center outside the PRP have generously shared their time and expertise with me through the years, including but not limited to Barbara Perry, Sheila Blackford, Douglas Blackmon, Andrew Chancey, Nancy Deane, Kim Curtis, Michael Greco, Rose Marie Owen, Russell Riley, and Kristy Schantz.

The University of Virginia Press under the bold leadership of Mark H. Saunders rose to the daunting technical, artistic, and literary challenge of publishing *Fatal Politics* both as a bound volume and as a linked ebook that provides readers with the opportunity to go straight from the text to the tapes and transcripts. Richard Holway, History and Social Sciences Editor at the Press, provided valuable and timely guidance. Managing Editor Ellen Satrom brought grace, patience, and élan to the challenge of shepherding the book to production. Copy Editor Susan Murray improved my prose with skill and flair. Cover designer Martha Farlow made us all look good. Marketing Director Jason Coleman and

Sales and Publicity Director Emily Grandstaff brought unflagging good cheer and skill to the art of connecting author and reader.

The Richard M. Nixon Library, under the pathfinding and courageous leadership of former Director Timothy Naftali, identified and made public thousands of invaluable documents illuminating a presidency that continues to cast a shadow on America in the twenty-first century. Even before the Nixon Library became a full and functional part of the National Archives and Records Administration, the Nixon Presidential Materials Project under Director Karl Weissenbach was extraordinarily helpful to a researcher finding his way through the vastest collection of documents generated by any presidency. Tapes Supervisory Archivist Cary McStay and the extraordinary team of archivists that has midwifed the lion's share of Nixon's tapes into the public domain deserve enormous credit for their achievement in a project unlike any other. More NARA archivists have helped me through the years than I can list here. They've demonstrated a level of skill and courtesy second to none. Most recently, Jon Fletcher broke speed records to supply candidate photos for the book cover.

The list of scholars who have lent me their help and expertise through the years grows longer than my memory, but it includes Stephen E. Ambrose, Taylor Branch, Brig. Gen. Charles F. Brower, Sahr Conway-Lanz, Mark Feldstein, Beverly Gage, Irwin F. Gellman, David Greenberg, George Herring, Stanley I. Kutler, Ralph Levering, Fredrik Logevall, Allen J. Matusow, Lien-Hang T. Nguyen, Chester Pach, Rick Perlstein, John Prados, Richard Reeves, Robert D. Schulzinger, Katherine Scott, Melvin Small, Evan Thomas, and Garry Wills.

Readers of *Fatal Politics* benefit indirectly from the great expertise and judgment of Vanderbilt University's Thomas A. Schwartz. Thanks to his thoughtful, probing, and incisive comments, the book is much improved from its raw first draft. Thanks to his generosity, thoughtfulness, and graciousness, the improvements were a pleasure to make. Marc Selverstone's keen eye for detail, vast knowledge of Cold War history, and persistence in revisiting issues helped me make more improvements to the first draft in both style and substance than I can count. It would be a different, and lesser, book without him.

Jeffrey Kimball, in addition to being the groundbreaking scholar of Nixon's "decent interval" exit strategy, was also big enough to raise no objection when I asked if we could use the cover picture from his landmark book, *Nixon's Vietnam War,* for the cover of this book as well. When it comes to the history of Nixon and Vietnam, Kimball is the giant on whose shoulders I stand.

Steve Coll, the *New Yorker* staff writer and Dean of Columbia Journalism School, and Eric Alterman of many platforms on- and offline provided early encouragement of my attempts to use the then-novel (for historians, anyway) medium of YouTube to disseminate evidence from the Nixon tapes to the wider public. Bob Woodward is still on the Nixon story, and his encouragement and support means a tremendous amount to someone who has literally spent years listening to the White House tapes. He and Carl Bernstein remain priceless sources of information and inspiration.

Family and friends make everything possible. Gerry Hughes gets why it's hilarious. Kirian Reddig gets why it's dreadful. Andrew Reddig knows how to get it done and generously does. Al Habjan is a wise and noble counselor. Tom Tully brings back Uncle Tony. Tim and Sarah Hughes make me proud, and their mother, Laurie, always impresses. Ernest and Mary Knobelspiesse, and their sons, Kirk and Eric, are kind, generous, and thoughtful—and therefore happy reminders of Alison. Deanna Linville is therapy for the mind and soul. Barbara Balestra has the brave soul of an explorer. Jeff Metzner and Jeffrey Eilender deserve medals for the years they spent listening to me talk about politics. Michael J. Hayes is the smartest man I know and possibly the wisest as well. Cosette Hayes is his wisest decision. Dessy Levinson is a valuable adviser and an invaluable friend. Too many people have been friends (and family by blood or spirit) to me over the years to mention here; thank you for making me wish I could.

The book's strengths reflect those of the above and all who have enriched my life; its flaws are mine.

Portions of this book have previously appeared in *Diplomatic History* and on the History News Network, History News Service, www.youtube.com/fatalpolitics, fatalpolitics.com, and millercenter.org.

NOTES

Introduction
1. Richard M. Nixon, *No More Vietnams* (New York: Arbor House, 1985), 9.
2. Ibid., 97, 145.
3. George W. Ball, "Block That Vietnam Myth," *New York Times*, 19 May 1985.
4. Nixon himself used the term "immoral" to describe the "decent interval" exit strategy when he claimed not to have adopted it. Nixon, *No More Vietnams*, 103.
5. Drew Westen, "Why Voters Say They Don't Really Know Barack Obama," *Huffington Post*, 6 September 2008, www.huffingtonpost.com/drew-westen/why-voters-say-they-dont_b_117238.html.

Fatal Politics
1. Louis Sims to Eugene P. Dagg, "Secret Service Participation in Tapings," 6 December 1973, "RG 87 Records of the U.S. Secret Service, Memoranda, CO-1-23206—WH Taping System 1971–1974, 4" folder, RG 87, Records of the U.S. Secret Service, Installation and Maintenance of the White House Sound Recording System and Tapes, CO-1-23206—WH Taping System . . . to Rm 522, Box 1, Richard M. Nixon Library (hereafter RMNL).
2. Alvin Snyder, "The Final Minutes," *TV Guide*, 6 August 1994, www.alvinsnyder.com/the_final_minutes_a_white_house_insider_s_intimate_look_at_a_vividly_historic_m_4636.htm.
3. Conversation 476-002, 7 April 1971, 8:59–9:20 p.m., Oval Office. All Nixon White House tapes come from the collection of the Nixon Presidential Library and are available online from the University of Virginia's Miller Center at www.millercenter.org or from www.nixonlibrary.org.
4. "Address to the Nation on the Situation in Southeast Asia," 7 April 1971, *Public Papers of the Presidents of the United States: Richard M. Nixon, 1971* (Washington, DC: GPO, 1972), www.presidency.ucsb.edu/ws/?pid=2972 (hereafter *PPPUS: Nixon, 1971*).
5. H. R. Haldeman, *The Haldeman Diaries: Inside the Nixon White House: The Complete Multimedia Edition* (Santa Monica, CA: Sony Electronic Publishing, 1994), 15 and 21 December 1970 (hereafter *Haldeman Diaries*); George Gallup, "73% Want Viet Pullout by Dec. 31," *Los Angeles Times*, 31 January 1971, www.proquest.com.
6. Kissinger was still calling Haldeman "Robert" in 2003, a decade after Haldeman's death (Henry Kissinger, *Ending the Vietnam War: A History of America's Involvement in and Extrication from the Vietnam War* [New York: Simon and Schuster, 2003], 63).

7. Haldeman's handwritten notes and memos to and from subordinates following up on presidential directives fill boxes at the Nixon Library in Yorba Linda, California (www.nixonlibrary.gov/forresearchers/find/textual/special/smof/haldeman.php). The Nixon Library also has copies of home movies shot in the then-popular Super 8 format by Haldeman and other Nixon aides (www.nixonlibrary.gov/forresearchers/find/av/motion_film/super_8.php).

8. H. R. Haldeman, "The Decision to Record Presidential Conversations," *Prologue Magazine,* Summer 1988, www.archives.gov/publications/prologue/1988/summer/haldeman.html.

9. *Haldeman Diaries,* 8. The historian Stephen E. Ambrose describes Haldeman's diary-recording practice in his introduction to the *Diaries.*

10. Ibid., 15 December 1970. Published version corrected with cassette copy of original tape-recorded entry at NARA, College Park, Maryland.

11. Ibid.

12. Egil Krogh, *The Day Elvis Met Nixon* (Bellevue, WA: Pejama Press, 1994), 9–12, 14, 31–37.

13. National Security Archive, "The Nixon-Presley Meeting 21 December 1970," www2.gwu.edu/~nsarchiv/nsa/elvis/elnix.html.

14. *Haldeman Diaries,* 21 December 1970.

15. Henry Kissinger, interview by Barbara Walters, *The Today Show,* NBC, aired 21 December 1970, White House Communications Agency Video Collection, WHCA VTR #4044.

16. "Address to the Nation on the Situation in Southeast Asia," 7 April 1971, *PPPUS: Nixon, 1971,* www.presidency.ucsb.edu/ws/?pid=2972.

17. Jim Stockdale and Sybil Stockdale, *In Love and War: The Story of a Family's Ordeal and Sacrifice during the Vietnam Years* (Annapolis, MD: Naval Institute Press, 1984), 302–4.

18. William Beecher, "Laird Appeals to Enemy to Release U.S. Captives," *New York Times,* 20 May 1969, www.proquest.com.

19. "Excerpts from Transcript of News Conference by Secretary of State Rogers," *New York Times,* 6 June 1969, www.proquest.com.

20. Michael J. Allen, *Until the Last Man Comes Home: POWs, MIAs, and the Unending Vietnam War* (Chapel Hill: University of North Carolina Press, 2009), Kindle edition, chap. 1, "Go Public: The Construction of Loss."

21. Philip Fradkin, "U.S. Team Reassuring Families of Viet POWs," *Los Angeles Times,* 11 July 1969, www.proquest.com.

22. Lael Morgan, "Doubt Rules Lives of POW Families," *Los Angeles Times,* 14 August 1969, www.proquest.com.

23. "Hanoi Officially Tells U.S. of POW Release," *Los Angeles Times,* 4 July 1969, www.proquest.com.

24. William Robbins, "Ex-P.O.W.'s Charge Hanoi with Torture," *New York Times,* 3 September 1969, www.proquest.com; John S. McCain III, "John McCain, Prisoner of War: A First-Person Account," *U.S. News and World Report,* 14 May 1973, www.usnews.com/news/articles/2008/01/28/john-mccain-prisoner-of-war-a-first-person-account.

25. "First Annual Report to the Congress on United States Foreign Policy for

the 1970's," 18 January 1970, *Public Papers of the Presidents of the United States: Richard M. Nixon, 1970* (Washington, DC: GPO, 1971), www.presidency.ucsb.edu/ws/?pid=2835 (hereafter *PPPUS: Nixon, 1970*).

26. Bryce Nelson, "List of 59 U.S. POWs in N. Vietnam Released," *Los Angeles Times*, 27 November 1969, www.proquest.com.

27. Tad Szulc, "Hanoi Radio Says U.S. Prisoners Can Receive Christmas Parcels," *New York Times*, 18 November 1969, www.proquest.com.

28. Conversation 471-2, 19 March 1971, 7:03–7:27 p.m., Oval Office.

29. Ibid.

30. "Address to the Nation on the Situation in Southeast Asia," 7 April 1971, *PPPUS: Nixon, 1971*, www.presidency.ucsb.edu/ws/?pid=2972.

31. Wayne Thompson, *To Hanoi and Back: The United States Air Force and North Vietnam, 1966–1973* (Washington, DC: Smithsonian Institution Press, 2000), Kindle edition, chap. 7, "Prisoners and Other Survivors."

32. "The President/Mr. Kissinger," 21 November 1970, 11:45 a.m., Kissinger Telcons Box 7, RMNL.

33. Richard M. Nixon, *RN: The Memoirs of Richard Nixon* (New York: Touchstone, 1978), 860.

34. Conversation 456-22, 23 February 1971, 4:12–6:18 p.m., Oval Office.

35. *Haldeman Diaries*, 23 November 1970.

36. Carol Mason, *Killing for Life: The Apocalyptic Narrative of Pro-Life Politics* (Ithaca, NY: Cornell University Press, 2002), 28.

37. "Address to the Nation on the Situation in Southeast Asia," 7 April 1971, *PPPUS: Nixon, 1971*, www.presidency.ucsb.edu/ws/?pid=2972.

38. Conversation 475-016, 8 April 1971, 9:18–10:07 a.m., Oval Office. Nixon, of course, was demanding more than the POWs, and Hanoi was demanding more than American withdrawal, but this conversation shows that Nixon realized that the release depended on the withdrawal.

39. Allen, *Until the Last Man Comes Home*, Kindle edition, chap. 1, "Go Public: The Construction of Loss."

40. "Text of Announcement by McGovern on 1972 Race," *New York Times*, 19 January 1971, www.nytimes.com.

41. Spencer Rich, "Senate Rejects End-War Plan by 55–39 Vote," *Washington Post*, 2 September 1970, www.proquest.com.

42. Conversation 465-008, 10 March 1971, 10:42 a.m.–1:15 p.m., Oval Office.

43. Conversation 456-005, 23 February 1971, 10:05–11:30 a.m., Oval Office.

44. "Address to the Nation on the War in Vietnam," 3 November 1969, *Public Papers of the Presidents of the United States: Richard M. Nixon, 1969* (Washington, DC: GPO, 1971), www.presidency.ucsb.edu/ws/?pid=2303 (hereafter *PPPUS: Nixon, 1969*).

Vietnamization

1. "Address to the Nation on the Situation in Southeast Asia," 7 April 1971, *PPPUS: Nixon, 1971*, www.presidency.ucsb.edu/ws/?pid=2972.

2. Ibid.

3. Louis Harris, "Public Backs Cambodia Step by Narrow 50–43% Margin,"

Washington Post, 25 May 1970, www.proquest.com; Louis Harris, "61% Now Believe Nixon Justified in Cambodia Move," *Washington Post*, 10 August 1970, www.proquest.com.

4. Henry Kissinger, *The White House Years* (London: Weidenfeld and Nicholson, 1979), 1004.

5. Conversation 457-001, 24 February 1971, 10:19–11:35 a.m., Oval Office.

6. Conversation 459-002, 27 February 1971, 9:35–11:57 a.m., Oval Office.

7. Conversation 464-012, 9 March 1971, 12:26–1:30 p.m., Oval Office.

8. "Address to the Nation on the Situation in Southeast Asia," 7 April 1971, *PPPUS: Nixon, 1971*, www.presidency.ucsb.edu/ws/?pid=2972.

9. Ibid.

10. E. W. Kenworthy, "Nixon Would Push for a Bigger Role by Saigon in War," *New York Times*, 30 September 1968, www.proquest.com; "Transcript of Speech by the Vice President on Foreign Policy," *New York Times*, 1 October 1968, www.proquest.com.

11. "Address to the Nation on the War in Vietnam," 3 November 1969, *PPPUS: Nixon, 1969*, www.presidency.ucsb.edu/ws/?pid=2303.

12. "The President's News Conference," 30 July 1970, *PPPUS: Nixon, 1970*, www.presidency.ucsb.edu/ws/?pid=2603.

13. NSC to Vice President's Office, "Revised Summary of Responses to NSSM-1: The Situation in Vietnam," 22 March 1969, Digital National Security Archive, http://gateway.proquest.com/openurl?url_ver=Z39.88-2004&res_dat=xri:dnsa&rft_dat=xri:dnsa:article:CPD01323.

14. "Address to the Nation on the War in Vietnam," 3 November 1969, *PPPUS: Nixon, 1969*, www.presidency.ucsb.edu/ws/?pid=2303.

15. "Address to the Nation on the Situation in Southeast Asia," 7 April 1971, *PPPUS: Nixon, 1971*, www.presidency.ucsb.edu/ws/?pid=2972.

16. Ibid.

17. Conversation 466-012, 11 March 1971, 4:00–4:55 p.m., Oval Office.

18. Ibid.

"A Nightmare of Recrimination"

1. "Address to the Nation on the Situation in Southeast Asia," 7 April 1971, *PPPUS: Nixon, 1971*, www.presidency.ucsb.edu/ws/?pid=2972.

2. Conversation 474-008, 26 March 1971, 4:09–4:53 p.m., Oval Office.

3. Conversation 457-001, 24 February 1971, 10:19–11:35 a.m., Oval Office.

4. "Address to the Nation on the Situation in Southeast Asia," 7 April 1971, *PPPUS: Nixon, 1971*, www.presidency.ucsb.edu/ws/?pid=2972.

5. Conversation 246-007, 7 April 1971, 12:16–2:00 p.m., Executive Office Building.

6. Bob Woodward and Carl Bernstein, *The Final Days* (New York: Simon and Schuster, 1976), Kindle edition, "Wednesday, August 7."

7. Conversation 001-010, 7 April 1971, 9:31–9:39 p.m., White House Telephone.

8. Conversation 246-017, 7 April 1971, 3:15–3:55 p.m., Executive Office Building.

9. Conversation 001-010, 7 April 1971, 9:31–9:39 p.m., White House Telephone.

"A Hell of a Shift"

1. Max Frankel, "Nixon Promises Pullout of 100,000 More G.I.'s by December," *New York Times;* Chalmers M. Roberts, "Nixon Sets 100,000 Troop Cut; President Refuses to Fix a Deadline," *Washington Post;* David Kraslow, "Vietnamization Success—Nixon." All three stories are from 8 April 1971, www.proquest.com.

2. James M. Naughton, "Salute Returned to Boy by Nixon; President Recalls Son of a Hero in Ending Speech," *New York Times,* 8 April 1971, www.proquest.com; Carl Bernstein, "Kevin Taylor, 4, Saluted Nixon February and Didn't Forget," *Washington Post,* 8 April 1971, www.proquest.com.

3. Summaries of *ABC Evening News* and *CBS Evening News,* 8 April 1971, http://tvnews.vanderbilt.edu.

4. Carl Bernstein, "Kevin Taylor, 4, Saluted Nixon February and Didn't Forget," *Washington Post,* 8 April 1971, www.proquest.com.

5. Conversation 475-016, 8 April 1971, 9:18–10:07 a.m., Oval Office.

6. Conversation 001-017, 7 April 1971, 10:07–10:16 p.m., White House Telephone.

7. Conversation 475-021, 8 April 1971, 1:12–2:00 p.m., Oval Office.

8. Conversation 460-028, 26 February 1971, 6:09–6:45 p.m., Oval Office; Brigadier General James D. Hughes to Haldeman, 2 March 1971, "General [James D.] Hughes March 1971" folder, Haldeman Box 75, Staff Member and Office Files—White House Special Files (hereafter WHSF-SMOF), RMNL.

9. Conversation 475-021, 8 April 1971, 1:12–2:00 p.m., Oval Office.

10. Ibid.

11. Conversation 476-014, 9 April 1971, 11:40 a.m.–1:30 p.m., Oval Office. Haldeman didn't name the pollster.

12. "3 Senators Aver Nixon Said He Had Pullout Deadline," *New York Times,* 9 April 1971, www.nytimes.com; David S. Broder and Spencer Rich, "White House, Scott Split on Pullout Date," *Washington Post,* 9 April 1971, www.proquest.com.

13. Conversation 476-007, 9 April 1971, 8:52–9:58 a.m., Oval Office.

How to Kill a Withdrawal Deadline

1. Conversation 479-007, 14 April 1971, 12:40–2:11 p.m., Oval Office.

2. Conversation 481-003, 17 April 1971, 8:56–9:13 a.m., Oval Office.

3. Ibid.

4. "Panel Interview at the Annual Convention of the American Society of Newspaper Editors," 16 April 1971, *PPPUS: Nixon, 1971,* www.presidency.ucsb.edu/ws/?pid=2982.

5. John Chancellor, *NBC Evening News,* aired 17 April 1971, White House Communications Agency Video Collection, WHCA #4300.

6. Don Oberdorfer, "President Links U.S. Withdrawal to POW Release; Capability of Saigon Also Stressed," *Washington Post,* 17 April 1971, www.proquest.com.

"I'm Being Perfectly Cynical"

1. "Address to the Nation on the War in Vietnam," 3 November 1969, *PPPUS: Nixon, 1969,* www.presidency.ucsb.edu/ws/?pid=2303.

2. Henry Kissinger, *Henry Kissinger: The Complete Memoirs E-book Boxed Set: "White House Years," "Years of Upheaval," "Years of Renewal"* (New York: Simon and Schuster, 2013), Kindle edition, chap. 12, "The War Widens."

3. "Address to the Nation on the Situation in Southeast Asia," 7 April 1971, *PPPUS: Nixon, 1971*, www.presidency.ucsb.edu/ws/?pid=2972.

4. Conversation 465-008, 10 March 1971, 10:42 a.m.–1:15 p.m., Oval Office.

5. Jeffrey Kimball, *The Vietnam War Files* (Lawrence: University Press of Kansas, 2004), 133–34. Nixon repeated his public demand that all outside forces withdraw when he revealed the secret talks in 1972 ("Address to the Nation Making Public a Plan for Peace in Vietnam," 25 January 1972, *PPPUS: Nixon, 1972*, www.presidency.ucsb.edu/ws/?pid=3475).

6. Conversation 507-004, 29 May 1971, 8:13–10:32 a.m., Oval Office.

7. William F. Buckley, *Right Reason* (Boston: Little, Brown, 1986), 311.

8. Conversation 507-004, 29 May 1971, 8:13–10:32 a.m., Oval Office.

9. Ibid.; "Memorandum of Conversation," 31 May 1971, 10:00 a.m.–1:30 p.m., "Camp David—Vol. VII" folder, NSCF Box 853, RMNL.

"We Want a Decent Interval"

1. "Polo I Kissinger (Briefing Book) July 1971 Trip to China," NSCF Box 850, RMNL.

2. Winston Lord, "Nixon in China, 40 Years Later," *Huffington Post*, 22 February 2012, www.huffingtonpost.com/winston-lord/nixon-in-china-40-years-l_b_1293643.html.

3. Conversation 002-052, 27 April 1971, 8:16–8:36 p.m., White House Telephone; "The President/Mr. Kissinger," 8:18 p.m., NSCF Box 1031, National Security Archive, RMNL, www2.gwu.edu/~nsarchiv/NSAEBB/NSAEBB66/ch-18.pdf.

4. Richard M. Nixon, "Asia after Viet Nam," *Foreign Affairs* (October 1967), www.foreignaffairs.com/articles/23927/richard-m-nixon/asia-after-viet-nam.

5. William Bundy, *A Tangled Web: The Making of Foreign Policy in the Nixon Presidency* (New York: Hill and Wang, 1998), 106.

6. Conversation 001-081, 14 April 1971, 7:27–7:40 p.m., White House Telephone.

7. Associated Press, "In '51, Nixon Criticized Truman Secrecy on Korea," *New York Times*, 24 June 1971, www.proquest.com.

8. Reagan found the perfect way to express both military-industrial supremacy and strategic frustration on this point: "It's silly talking about how many years we will have to spend in the jungles of Vietnam when we could pave the whole country and put parking stripes on it and still be home by Christmas." Paving North Vietnam, however, would have given the Chinese a highway south (Rick Perlstein, *The Invisible Bridge: The Fall of Nixon and the Rise of Reagan* [New York: Simon and Schuster, 2014], Kindle edition, chap. 5, "A Whale of a Good Cheerleader").

9. Reuters, "China Won't Be Idle; Hanoi Warns Nixon," *Washington Post*, 19 February 1971, www.proquest.com.

10. Conversation 451-023, 18 February 1971, 6:16–6:37 p.m., Oval Office.

11. "Memorandum of Conversation," 9 July 1971, 4:35–11:20 p.m., attached to Lord to Kissinger, "Memcon of Your Conversations with Chou En-lai," 29 July 1971, NSCF Box 1033, RMNL.
12. Conversation 001-091, 14 April 1971, 8:05–8:12 p.m., White House Telephone.
13. Conversation 001-101, 15 April 1971, 7:31–7:33 p.m., White House Telephone.
14. Conversation 534-002, 1 July 1971, 8:45–9:52 a.m., Oval Office.
15. Conversation 532-011, 30 June 1971, 10:18–10:30 a.m., Oval Office.
16. Conversation 534-003, 1 July 1971, 9:54–10:26 a.m., Oval Office.
17. Conversation 534-002, 1 July 1971, 8:45–9:52 a.m., Oval Office.
18. Conversation 534-003, 1 July 1971, 9:54–10:26 a.m., Oval Office.

Meeting Zhou

1. Kissinger, *White House Years,* 743.
2. Ibid., 743, 745.
3. Ibid., 746.
4. "Memorandum of Conversation," 9 July 1971, 4:35–11:20 p.m., attached to Lord to Kissinger, "Memcon of Your Conversations with Chou En-lai," 29 July 1971, NSCF Box 1033, RMNL.
5. Kissinger, *White House Years,* 743–49.
6. "Memorandum of Conversation," 9 July 1971, 4:35–11:20 p.m., attached to Lord to Kissinger, "Memcon of Your Conversations with Chou En-lai," 29 July 1971, NSCF Box 1033, RMNL.
7. Jussi Hanhimäki deserves credit for first uncovering most of the quotes from Polo I in this chapter (Hanhimäki, "Some More 'Smoking Guns'? The Vietnam War and Kissinger's Summitry with Moscow and Beijing, 1971–1973," *Passport,* December 2001).

"Old Friends"

1. "Remarks to the Nation Announcing Acceptance of an Invitation to Visit the People's Republic of China," 15 July 1971, *PPPUS: Nixon, 1971,* www.presidency.ucsb.edu/ws/?pid=3079.
2. John Chancellor, "Richard Nixon Announcement Re: Trip to China," NBC Special Program, 15 July 1971, http://tvnews.vanderbilt.edu; Seymour Topping, "'Journey for Peace'; President Will Seek to Break Down 21-Year-Old Great Wall of Hostility," *New York Times,* 16 July 1971, www.proquest.com; David Kraslow, "Profound Impact on Foreign Affairs Expected," *Los Angeles Times,* 16 July 1971, www.proquest.com; Frank Starr and John W. Finney, "Congress Chiefs Pleased; Support Is Bipartisan," *New York Times,* 16 July 1971, www.proquest.com; Associated Press, "NBC Bids for Live TV on Nixon in China," *Chicago Tribune,* 16 July 1971, www.proquest.com.
3. Sezig S. Harrison, "Taiwan Protests Nixon Announcement," *Washington Post,* 16 July 1971, www.proquest.com; William Fulton, "17 Nations Ask U.N. to Seat Red China," *Chicago Tribune,* 16 July 1971, www.proquest.com; Tribune

Wire Service, "Ford Sees Indochina Peace Talks Resulting from Visit," *Chicago Tribune,* 16 July 1971, www.proquest.com.

4. Carroll Kilpatrick, "Groundwork Laid by Kissinger, Chou in Secret Meeting," *Washington Post,* 16 July 1971, www.proquest.com; Tad Szulc, "Kissinger Visit Capped 2-Year Effort," *New York Times,* 16 July 1971, www.proquest.com; Chicago Tribune Press Service, "Kissinger Sure to Be a Legend: Already a Swinger," *Chicago Tribune,* 16 July 1971, www.proquest.com; *Time* cover, 26 July 1971, www.timecoverstore.com/product/henry-kissinger-and-richard-nixon-1971-07-26; Bernard Law Collier, "The Road to Peking, or, How Does This Kissinger Do It?," *New York Times,* 14 November 1971, www.proquest.com.

"He Deserves Our Confidence"

1. United Press International, "Nixon Move Pleases Reagan," *New York Times,* 17 July 1971, www.proquest.com.

2. "New Policy for China Imperative, Says Nixon," *Los Angeles Times,* 7 November 1950, www.proquest.com.

3. The House has a handy chart of membership by party throughout its history at http://history.house.gov/Institution/Party-Divisions/Party-Divisions/. Robert C. Albright, "Senate Count, 49-97, Aids Coalition Power; GOP Boosts House Membership by 31," *Washington Post,* 9 November 1950, www.proquest.com.

4. Carl Greenberg, "Nixon's Convention Delegation Open to All, Reagan Says," *Los Angeles Times,* 3 October 1971, www.proquest.com.

5. Ibid.

6. "The Two Worlds: A Day-Long Debate," *New York Times,* 25 July 1959, www.proquest.com.

7. "Address before a Joint Session of Congress," 25 January 1984, *Public Papers of the Presidents of the United States: Ronald Reagan, 1984* (Washington, DC: GPO, 1986), www.reagan.utexas.edu/archives/speeches/1984/12584e.htm.

8. Carl Greenberg, "Nixon's Convention Delegation Open to All, Reagan Says," *Los Angeles Times,* 3 October 1971, www.proquest.com.

9. Consider Nixon's attacks on Adlai Stevenson, the 1952 Democratic presidential nominee. Nixon called him "Adlai the appeaser . . . who got a Ph.D. degree from [Secretary of State Dean] Acheson's College of Cowardly Communist Containment" (David Greenberg, *Nixon's Shadow: The History of an Image* [New York: Norton, 2004], 55). While claiming not to question the Democrat's loyalty, Nixon said, "Mr. Stevenson has lined up consistently with those who minimize and cover up the Communist threat" (Associated Press, "Nixon Declares Adlai 'Color Blind' on Reds," *Washington Post,* 26 October 1952, www.proquest.com).

JFK v. Nixon

1. Arthur M. Schlesinger Jr., *A Thousand Days: John F. Kennedy in the White House* (Boston: Houghton Mifflin, 1965), Kindle edition, chap. 9, "The Hour of Euphoria: Castro and Kennedy."

2. "Text of Statement by Kennedy on Dealing with Castro Regime," *New York Times,* 21 October 1960, www.proquest.com.

3. Peter Kihss, "Kennedy Asks Aid for Cuban Rebels to Defeat Castro," *New York Times*, 21 October 1960, www.proquest.com.

4. Jack Raymond, "Pentagon Backed by Key Democrat," *New York Times*, 4 May 1960, www.proquest.com; Schlesinger, *A Thousand Days*, Kindle edition, chap. 12, "New Departures: The Reconstruction of National Strategy"; Robert S. McNamara and Brian VanDeMark, *In Retrospect: The Tragedy and Lesson of Vietnam* (New York: Times Books, 1995), 20.

5. Jack Raymond, "President Backs Defense Program as Criticism Rises," *New York Times*, 28 August 1958, www.proquest.com.

6. Jack Raymond, "Kennedy Defense Study Finds No Evidence of a 'Missile Gap,'" *New York Times*, 7 February 1961, www.proquest.com; McNamara and VanDeMark, *In Retrospect*, 21.

7. "September 26, 1960 Debate Transcript," Commission on Presidential Debates, www.debates.org/index.php?page=september-26-1960-debate-transcript. Kennedy misquoted Lincoln, who asked if America could endure "half slave *and* half free." Asking if the nation or world could exist "half-slave *or* half-free" is like asking if a glass of milk is half-full or half-empty.

8. "Eisenhower Gives Nixon His Full Backing, Says He Took Part in Many Key Decisions," *Wall Street Journal*, 30 September 1960, www.proquest.com.

9. Richard M. Nixon, "Nixon Calls Kennedy's Foreign Policy Weak," *Washington Post*, 24 July 1961, www.proquest.com.

10. United Press International, "Nixon Says Kennedy Erred in Cuba, Laos," *Los Angeles Times*, 14 September 1961, www.proquest.com.

11. Don Shannon, "Macmillan, Kennedy OK Laos Plan," *Los Angeles Times*, 7 April 1961, www.proquest.com.

12. Richard M. Nixon, "Nixon Calls Kennedy's Foreign Policy Weak," *Washington Post*, 24 July 1961, www.proquest.com.

13. Reuters, "Coalition Regime in Laos Abolished," *New York Times*, 4 December 1975, www.proquest.com.

14. United Press International, "Nixon Backs Kennedy Build-Up of U.S. Armed Force in Vietnam," *New York Times*, 16 February 1962, www.proquest.com.

15. Daniel Ellsberg, *Secrets: A Memoir of Vietnam and the Pentagon Papers* (New York: Viking, 2002), 195.

16. "I didn't say I was going to force a coalition government on South Vietnam," McCarthy replied ("Excerpts from the Kennedy-McCarthy Televised Discussion," *New York Times*, 2 June 1968, www.proquest.com; R. W. Apple Jr., "Kennedy Disputes M'Carthy on War in TV Discussion," *New York Times*, 2 June 1968, www.proquest.com; Ellsberg, *Secrets*, 195).

17. Cong. Rec. 18,783 (8 June 1971) (qtd. in statement of Sen. Philip Hart).

18. R. W. Apple Jr., "Nixon Would Bar Forced Coalition in South Vietnam," *New York Times*, 28 October 1968, www.nytimes.com.

19. "Address to the Nation on the War in Vietnam," 3 November 1969, *PPPUS: Nixon, 1969*, www.presidency.ucsb.edu/ws/?pid=2303.

The Kennedy Critique
1. Cong. Rec. 18,783 (8 June 1971) (qtd. in statement of Sen. Philip Hart).
2. Kenneth O'Donnell, "LBJ and the Kennedys," *Life*, 7 August 1970.
3. Debate about JFK's intentions regarding Vietnam rages on. Some of it is captured by James Blight, Janet M. Lang, and David A. Welch in *Vietnam If Kennedy Had Lived: Virtual JFK* (Lanham, MD: Rowman and Littlefield, 2009).
4. Kenneth P. O'Donnell, David F. Powers, and Joe McCarthy, *Johnny, We Hardly Knew Ye: Memories of John Fitzgerald Kennedy* (Boston: Little, Brown, 1970, 1972), 18.
5. Ellsberg, *Secrets*, 195.
6. Rusk is quoted in Michael Charlton, Anthony Charlton, and Anthony Moncrieff, *Many Reasons Why* (New York: Hill and Wang, 1978), 82. See also Dean Rusk and Richard Rusk, *As I Saw It*, ed. Daniel S. Papp (New York: Norton, 1990), 441–42.
7. Ronald Reagan, untitled speech at Republican fund-raising dinner in Boston, Massachusetts, 14 June 1971, Tape #432, Gubernatorial Audiotape Collection, Ronald Reagan Presidential Library; David Kraslow, "Reagan Scores Kennedy for Stand on War," *Los Angeles Times*, 15 June 1971, www.proquest.com.
8. In the 1968 presidential campaign, Sen. Bourke Hickenlooper, R-Iowa, accused Lyndon Johnson of halting the bombing of North Vietnam to sway the election in favor of Vice President Hubert Humphrey, saying, "It's tragic that American lives are being played with this way." John W. Finney, "Doves and Hawks Divided on Johnson's Move," *New York Times*, 1 November 1968, www.nytimes.com. For a thorough examination of LBJ's motives for halting the bombing, see Ken Hughes, *Chasing Shadows: The Nixon Tapes, The Chennault Affair, and the Origins of Watergate* (Charlottesville: University of Virginia Press, 2014), 1–43. More than one Republican accused Franklin Roosevelt of advance knowledge of the Japanese attack on Pearl Harbor. Although LBJ and FDR were thoroughly political men, these specific allegations of playing politics with American lives were false—unlike the one that Kennedy was leveling against Nixon. Nevertheless, Republicans cheered Reagan's paean to the nobility of their party's politicians (David Greenberg, "Who Lost Pearl Harbor?" *Slate*, 7 December 2000, www.slate.com/articles/news_and_politics/history_lesson/2000/12/who_lost_pearl_harbor.html).
9. David Kraslow, "Apologize to President, Reagan Tells Kennedy," *Los Angeles Times*, 16 June 1971, www.proquest.com.
10. United Press International, "Humphrey Says Nixon Shuns Politics on War," *New York Times*, 10 June 1971, www.proquest.com.
11. Cong. Rec. 18,936 (9 June 1971) (statement of Sen. Hubert Humphrey).
12. United Press International, "Humphrey Says Nixon Shuns Politics on War," *New York Times*, 10 June 1971, www.proquest.com.
13. Conversation 005-002, 10 June 1971, 2:53–2:57 p.m., White House Telephone.
14. Ibid.
15. "Transcript of Speech by the Vice President," *New York Times*, 1 October 1968, www.proquest.com.

16. It's interesting to note that Humphrey said he, too, would have appointed Henry Kissinger national security adviser if he, not Nixon, had won the 1968 election. With the same adviser, President Humphrey would have received much the same advice. Hubert H. Humphrey, *The Education of a Public Man: My Life and Politics*, ed. Norman Sherman (Garden City, NY: Doubleday, 1976), 9.

17. Mark Katz has made the phrase "leaving without losing" famous with a book of that title. I independently came up with the same alliteration and used it publicly before his book came out, but Katz deserves credit for coining it as well and for getting it into print first (see Mark N. Katz, *Leaving without Losing: The War on Terror after Iraq and Afghanistan* [Baltimore: Johns Hopkins University Press, 2012]).

18. "Address to the Nation on Vietnam," 14 May 1969, *PPPUS: Nixon, 1969*, www.presidency.ucsb.edu/ws/?pid=2047; "Nixon's Popularity Found at High Point," *New York Times*, 6 February 1973, www.proquest.com.

19. David R. Derge, Vice President and Dean of Indiana University, to President Nixon, "The Public Appraises the Nixon Administration and Key Issues (With Particular Emphasis on Vietnam)," 11 August 1969, "E.O.B. Office Desk—August 10, 1974" folder, Box 185, President's Personal File, Materials Removed from President's Desk, 1969–74, [EOB Office Desk . . . Administration] to [Blank Stationery— . . . August 9, 1974], RMNL.

20. "Address to the Nation on the War in Vietnam," 3 November 1969, *PPPUS: Nixon, 1969*, www.presidency.ucsb.edu/ws/?pid=2303.

21. George Gallup, "Nixon's Popularity Rises to New High," *Los Angeles Times*, 23 November 1969, www.proquest.com.

22. "Nixon's Popularity Found at High Point," *New York Times*, 6 February 1973, www.proquest.com.

23. "Address to the Nation on the Situation in Southeast Asia," 30 April 1970, *PPPUS: Nixon, 1970*, www.presidency.ucsb.edu/ws/?pid=2490.

24. "Address to the Nation on the Situation in Southeast Asia," 8 May 1972, *Public Papers of the Presidents of the United States: Richard M. Nixon, 1972* (Washington, DC: GPO, 1974), www.presidency.ucsb.edu/ws/?pid=3404 (hereafter *PPPUS: Nixon, 1972*).

The Liberal Mistake

1. Hughes, *Chasing Shadows*, 94–96.

2. Hoover to Nixon, 29 December 1969, and undated handwritten notes between Ehrlichman and Haldeman, in House Committee on the Judiciary, *Statement of Information: Book 7, Part 1: White House Surveillance Activities and Campaign Activities* (Washington, DC: GPO, 1974), 359–68.

3. Halperin affidavit, 30 November 1973, in House Committee on the Judiciary, *Statement of Information: Book 7, Part 1: White House Surveillance Activities and Campaign Activities*, 218–21.

4. "Address to the Nation on the War in Vietnam," 3 November 1969, *PPPUS: Nixon, 1969*, www.presidency.ucsb.edu/ws/?pid=2303; "Address to the Nation on Progress toward Peace in Vietnam," 15 December 1969, ibid., www.presidency.ucsb.edu/ws/?pid=2370; "Address to the Nation on Progress toward Peace in

Vietnam," 20 April 1970, *PPUS: Nixon, 1970,* www.presidency.ucsb.edu/ws
/?pid=2476.

5. Les Gelb and Morton H. Halperin, "Only a Timetable Can Extricate Nixon," *Washington Post,* 24 May 1970, www.proquest.com.

6. Leslie H. Gelb and Morton H. Halperin, "Two 'Offers' on Vietnam Are Still Far Apart," *Washington Post,* 11 October 1970, www.proquest.com.

7. Townsend Hoopes and Paul C. Warnke, "Nixon Is Really Just Digging In," *Washington Post,* 21 June 1970, www.proquest.com.

8. See John Herbers, "Clifford Terms War Key '72 Issue," *New York Times,* 23 June 1972, www.proquest.com.

9. "Reaction to Talk Mixed in Capitol," *New York Times,* 8 April 1971, www.proquest.com.

10. "Excerpts from Democrats' Remarks on Vietnam," *New York Times,* 23 April 1971, www.proquest.com.

11. Ellsberg, *Secrets,* 229.

12. Ibid., 257–58.

"Super Secret Agent"

1. Richard Reeves, *Alone in the White House* (New York: Simon and Schuster, 2001), 428.

2. "Memorandum of Conversation," 22 February 1972, 2:10–6:10 p.m., President's Office Files Box 87, RMNL.

3. Ibid.

Sixty-Six Percent for Six Months

1. "Address to the Nation on the Situation in Southeast Asia," 8 May 1972, *PPPUS: Nixon, 1972,* www.presidency.ucsb.edu/ws/?pid=3404.

2. Nixon, *No More Vietnams,* 145.

3. David R. Derge, Vice President and Dean of Indiana University, to President Nixon, "The Public Appraises the Nixon Administration and Key Issues (With Particular Emphasis on Vietnam)," 11 August 1969, "E.O.B. Office Desk—August 10, 1974" folder, Box 185, President's Personal File, Materials Removed from President's Desk, 1969–74, [EOB Office Desk . . . Administration] to [Blank Stationery— . . . August 9, 1974], RMNL.

4. Nixon, *RN,* 603.

5. "Address to the Nation on the Situation in Southeast Asia," 8 May 1972, *PPPUS: Nixon, 1972,* www.presidency.ucsb.edu/ws/?pid=3404.

6. Craig R. Whitney, "Foe Sweeps across DMZ; Saigon Troops Fall Back; Clouds Block U.S. Plans; Advisers Uneasy," *New York Times,* 2 April 1972, www.proquest.com; Craig Whitney, "Half of Province in South Vietnam Lost to Invaders," *New York Times,* 3 April 1972, www.proquest.com; "N. Viet Invasion Stalls as Allies Launch Huge Counteroffensive," *Los Angeles Times,* 4 April 1972, www.proquest.com; Associated Press, "Reds Open Front near Saigon; South Vietnam Fights for Life, Thieu Says," *Los Angeles Times,* 5 April 1972, www.proquest.com; Michael Getler, "U.S. Sternly Warns Hanoi, Readies New Air Buildup," *Washington Post,* 7 April 1972, www.proquest.com; "Laird Con-

firms U.S. Will Bomb until Reds Withdraw, Negotiate," *Los Angeles Times*, 7 April 1972, www.proquest.com.

7. Craig R. Whitney, "U.S. Analysts in Saigon Say Hanoi Threw All but One Division into the Offensive," *New York Times*, 10 April 1972, www.proquest.com. Hanoi had two additional "training" divisions in the North (Associated Press, "U.S. Identifies Hanoi Divisions in Fighting," *Los Angeles Times*, 22 April 1972, www.proquest.com). George McArthur, "100,000 Enemy Force Seen in Action Soon," *Los Angeles Times*, 10 April 1972, www.proquest.com.

8. Malcolm W. Browne, "Key Highlands Base Reported Overrun in a Major Offensive by Enemy Tanks," *New York Times*, 24 April 1972, www.proquest.com; Craig R. Whitney, "Saigon's Forces Flee in Disorder toward Kontum," *New York Times*, 25 April 1972, www.proquest.com.

9. William M. Hammond, *Public Affairs: The Military and the Media, 1968–1973*, United States Army in Vietnam series (Washington, DC: Center of Military History, United States Army, 1996), 538.

10. Reuters, "U.S. War Dead at Six-Month High," *Los Angeles Times*, 7 April 1972, www.proquest.com.

11. Hammond, *Public Affairs, 1968–1973*, 538–39.

12. "Address to the Nation on Vietnam," 26 April 1972, *PPPUS: Nixon, 1972*, www.presidency.ucsb.edu/ws/?pid=3384.

13. Alexander M. Haig Jr., *Inner Circles: How America Changed the World: A Memoir* (New York: Warner, 1992), 282.

14. Malcolm W. Browne, "Fear of Foe Grips People of Pleiku; Hundreds Try to Flee Town in Highlands Expecting the Enemy to Overrun It Soon," *New York Times*, 29 April 1972, www.proquest.com; Fox Butterfield, "Enemy Artillery Batters Quangtri as Ring Tightens," *New York Times*, 30 April 1972, www.proquest.com; Malcolm W. Browne, "Thousands Flee Kontum in Panic as Enemy Nears," *New York Times*, 1 May 1972, www.proquest.com; Peter Osnos, "Quangtri's Fall Stuns South Vietnam; Loss of Quangtri Province Shakes Vietnam's Morale," *Washington Post*, 3 May 1972, www.proquest.com; Sydney Schanberg, "'It's Everyone for Himself' As Troops Rampage in Hue," *New York Times*, 4 May 1972, www.proquest.com; Fox Butterfield, "Enemy Overruns Base near Pleiku, Killing about 80," *New York Times*, 6 May 1972, www.proquest.com.

15. "Address to the Nation on the Situation in Southeast Asia," 8 May 1972, *PPPUS: Nixon, 1972*, www.presidency.ucsb.edu/ws/?pid=3404.

16. Nixon, *No More Vietnams*, 145.

17. Conversation 334-010, 2 May 1972, 9:28–10:00 a.m., Executive Office Building.

18. Qtd. in Hammond, *Public Affairs, 1968–1973*, 567.

19. Conversation 726-001, 19 May 1972, 10:30–11:42 a.m., Oval Office.

20. Hammond, *Public Affairs, 1968–1973*, 568.

21. Conversation 335-033, 5 May 1972, 2:10–3:15 p.m., Executive Office Building.

22. Lewis Sorley, *A Better War: The Unexamined Victories and Final Tragedy of America's Last Years in Vietnam* (San Diego: Harcourt, 1999), 327–28.

23. Lewis Sorley, *Vietnam Chronicles: The Abrams Tapes, 1968–1972* (Lubbock: Texas Tech University Press, 2004), 833.

24. Terence Smith, "'Test Has Finally Come': Administration Officials Are Cautious but Optimistic on Vietnamization Fate," *New York Times*, 6 April 1972, www.proquest.com.

25. Sorley, *A Better War*, 329–30.

26. Ibid., 325.

27. Hammond, *Public Affairs, 1968–1973*, 564–66.

28. Ibid., 567–68.

29. John Randolph, "U.S. Leaving Khe Sanh Because It Became a Military Liability," *Los Angeles Times*, 30 June 1968, www.proquest.com.

30. Ibid.

31. Nixon, *No More Vietnams*, 150–51. The "debacle" quote of Abrams comes from Hammond, *Public Affairs, 1968–1973*, 553. The Office of Joint History agreed that "it should have been equally clear that if Saigon's forces alone had been pitted against Hanoi's, the South Vietnamese would not have fought successfully" (Willard J. Webb and Walter S. Poole, *The Joint Chiefs of Staff and the War in Vietnam, 1971–1973* [Washington, DC: GPO, 2007], 160).

32. Stephen P. Randolph, *Powerful and Brutal Weapons: Nixon, Kissinger, and the Easter Offensive* (Cambridge: Harvard University Press, 2007), 337.

33. Sorley, *A Better War*, 327; Webb and Poole, *The Joint Chiefs of Staff and the War in Vietnam, 1971–1973*, 177; Randolph, *Powerful and Brutal Weapons*, 338; Nixon, *No More Vietnams*, 148–49; Kimball, *The Vietnam War Files*, 238. Kimball quotes an 11 August 1972 CIA memo assessing the bombing and mining's impact.

34. Nixon, *No More Vietnams*, 149.

35. Webb and Poole, *The Joint Chiefs of Staff and the War in Vietnam, 1971–1973*, 187.

36. Tad Szulc, "Hanoi Held Able to Fight 2 Years at Present Rate," *New York Times*, 13 September 1972, www.proquest.com.

37. CIA Intelligence Memorandum, 11 August 1972, qtd. in Kimball, *The Vietnam War Files*, 238; *Foreign Relations of the United States (FRUS), 1969–1976: Vietnam, January–October 1972*, ed. John M. Carland (Washington, DC: GPO, 2010), 8: Document 236 (hereafter *FRUS 1969–1976*, 8).

38. Stephen Randolph writes that Linebacker was indeed "crippling" the North, yet he finds that despite this, the North's troops managed to stay in the South "and ultimately to achieve their minimum strategic objectives" (Randolph, *Powerful and Brutal Weapons*, 338).

39. Laird to Kissinger, 6 April 1972, Digital National Security Archive, http://gateway.proquest.com/openurl?url_ver=Z39.88-2004&res_dat=xri:dnsa&rft_dat=xri:dnsa:article:CVW00105.

40. Nixon, *No More Vietnams*, 145.

41. Ellsberg, *Secrets*, 416.

42. Spencer Rich, "A Kissinger Study," *Washington Post*, 25 April 1972, www.proquest.com.

43. R. J. Smith to Kissinger, "NSC Staff Review of Response to NSSM 1," 19 March 1969, Digital National Security Archive, http://gateway.proquest.com/openurl?url_ver=Z39.88-2004&res_dat=xri:dnsa&rft_dat=xri:dnsa:article:CPR00283.

44. Tad Szulc, "1969 Study Shows War Policy Split," *New York Times*, 26 April 1972, www.proquest.com. Only latter-day muckraker Jack Anderson focused on NSSM-1's implications for Vietnamization: "All the experts agreed that the South Vietnamese armed forces, 'in the foreseeable future,' couldn't fight off the Vietcong and North Vietnamese 'without U.S. combat support in the form of air, helicopters, logistics and some ground forces'" (Jack Anderson, "'69 Study Told of Saigon Weakness," *Washington Post*, 26 April 1972, www.proquest.com). Ellsberg didn't have the final version of NSSM-1, which put Saigon's dependency in stronger terms by saying the South needed the United States to provide "major ground forces" (NSC to Vice President's Office, "Revised Summary of Responses to NSSM-1: The Situation in Vietnam," 22 March 1969, Digital National Security Archive, http://gateway.proquest.com/openurl?url_ver=Z39.88-2004&res_dat =xri:dnsa&rft_dat=xri:dnsa:article:CPD01323).

45. Nixon, *No More Vietnams*, 146.

46. Merle L. Pribbenow, trans., *Victory in Vietnam: The Official History of the People's Army of Vietnam, 1954–1975* (Lawrence: University Press of Kansas, 2002), 301.

47. Nixon, *No More Vietnams*, 150.

48. Louis Harris, "59% of Public Backs Nixon Viet Moves," *Washington Post*, 13 May 1972, www.proquest.com.

49. Conversation 726-001, 19 May 1972, 10:30–11:42 a.m., Oval Office.

50. Conversation 773-012, 8 September 1972, 11:49 a.m.–12:12 p.m., Oval Office.

51. James Mann, *The Rebellion of Ronald Reagan: A History of the End of the Cold War* (New York: Viking, 2009), 233.

"One Arm Tied Behind"

1. Earl Mazo, "Nixon Asserts U.S. Risks Defeat Soon in Vietnam Conflict," *New York Times*, 27 January 1965, www.proquest.com; Nixon, *RN*, 270–71.

2. "Goldwater Asks More Air Strikes," *New York Times*, 22 February 1965, www.nytimes.com.

3. James Reston, "Washington: The Politics of Vietnam," *New York Times*, 16 June 1965, www.nytimes.com.

4. E. W. Kenworthy, "G.O.P. House Chief in Vietnam Plea," *New York Times*, 2 July 1965, www.nytimes.com.

5. David S. Broder, "G.O.P. Finds Rising Peril of 'Endless' Vietnam War," *New York Times*, 14 December 1965, www.nytimes.com.

6. 112-a Cong. Rec. 6404 (daily ed., 21 March 1966) (statement of Sen. Richard Russell).

7. "How We Can Win in Viet Nam," *Human Events*, 28 January 1967, www.proquest.com.

8. "Reagan Urges Escalation to Win the War 'Quickly,'" *New York Times*, 13 September 1967, www.proquest.com.

9. Qtd. in Bernard von Bothmer, *Framing the Sixties: The Use and Abuse of a Decade from Ronald Reagan to George W. Bush* (Amherst: University of Massachusetts Press, 2010), Kindle edition, chap. 4, "Reagan and the Memory of the Vietnam War: Purging the Ghosts of Vietnam."

"Why Does the Air Force Constantly Undercut Us?"

1. Conversation 335-033, 5 May 1972, 2:10–3:15 p.m., Executive Office Building.
2. Conversation 726-001, 19 May 1972, 10:30–11:42 a.m., Oval Office.

The Appearance of Success

1. Webb and Poole, *The Joint Chiefs of Staff and the War in Vietnam, 1971–1973*, 164–65.
2. Jacques Leslie, "S. Vietnam Troops Turn Back Red Attack at Kontum," *Los Angeles Times*, 15 May 1972, www.proquest.com; George McArthur, "S. Viet Units Break Long Kontum Siege, Push on Quang Tri," *Los Angeles Times*, 1 July 1972, www.proquest.com; Joseph B. Treaster, "Saigon Reports Its Troops Enter Quangtri Capital," *New York Times*, 5 July 1972, www.proquest.com; Associated Press, "Saigon Reports Repulsing an Enemy Attack on a Pleiku Outpost," *New York Times*, 6 September 1972, www.proquest.com.
3. "I will only say the bombing and mining was essential to turn around what was a potentially disastrous situation in South Vietnam. The back of the enemy offensive has been broken. They hold no provincial capitals now at all. This could not have been accomplished without the mining and the bombing," Nixon said ("The President's News Conference," 5 October 1972, *PPPUS: Nixon, 1972*, www.presidency.ucsb.edu/ws/?pid=3617).
4. Merriam-Webster's online dictionary translates it from the Latin as "after this, therefore on account of it (a fallacy of argument)" (www.merriam-webster.com/dictionary/post-hoc-ergo-propter-hoc).

"Any Means Necessary"

1. Nixon, *No More Vietnams*, 149.
2. Larry Berman, *No Peace, No Honor: Nixon, Kissinger, and Betrayal in Vietnam* (New York: Free Press, 2001), 37–39.
3. Unsigned and undated document in "Sensitive Camp David—Vol. 1" folder, NSCF Box 852, RMNL. The document is unidentifiable as the ultimatum Sainteny issued on Nixon's behalf because Nixon quoted the "to measures of great consequence and force" passage in his memoirs, but not the more threatening "by any means necessary" passage (Nixon, *RN*, 393–94).
4. Rowland Evans and Robert Novak, "'Nixon Is Talking about the War Just Like Johnson Used to Talk,'" *Washington Post*, 8 October 1969, www.proquest.com; Nixon, *RN*, 400.
5. Nixon and Eisenhower are quoted in Jeffrey Kimball, *Nixon's Vietnam War* (Lawrence: University Press of Kansas, 1998), 82–83.
6. William Burr and Jeffrey Kimball, "Nixon's Nuclear Ploy," *Bulletin of Atomic Scientists* 59, no. 1 (January 2003): 28–37, 72–73, http://bos.sagepub.com/content/59/1/28. Burr and Kimball will dig deeper into Nixon's secret nuclear alert in *Nixon's Nuclear Specter: The 1969 Secret Alert, Madman Diplomacy, and the Vietnam War* (Lawrence: University Press of Kansas, forthcoming).
7. Nixon, *RN*, 401.
8. Ibid., 402.

9. David R. Derge, Vice President and Dean of Indiana University, to President Nixon, "The Public Appraises the Nixon Administration and Key Issues (With Particular Emphasis on Vietnam)," 11 August 1969, "E.O.B. Office Desk—August 10, 1974" folder, Box 185, President's Personal File, Materials Removed from President's Desk, 1969–74, [EOB Office Desk . . . Administration] to [Blank Stationery— . . . August 9, 1974], RMNL.

10. Nixon, *RN*, 402.

11. Lien-Hang T. Nguyen writes that Ho "might have been amenable had he not been marginalized in the Politburo and nearing the end of his life," but Nixon's offer was rejected by Le Duan, general secretary of the Central Committee of North Vietnam's Communist Party (*Hanoi's War: An International History of the War for Peace in Vietnam* [Chapel Hill: University of North Carolina Press, 2012], Kindle edition, chap. 4, "To Paris and Beyond: The Nixon Doctrine and Kissingerian Diplomacy").

12. Nixon, *RN*, 404–5.

13. Kissinger to Nixon, "Contingency Military Operations against North Vietnam," 2 October 1969, "Top Secret/Sensitive Vietnam Contingency Planning, Oct. 2, 1969 [2 of 2]" folder, NSCF Box 89, RMNL.

14. Qtd. in Kimball, *The Vietnam War Files*, 105.

15. "Address to the Nation on the War in Vietnam," 3 November 1969, *PPPUS: Nixon, 1969*, www.presidency.ucsb.edu/ws/?pid=2303.

"A Russian Game, a Chinese Game and an Election Game"

1. Nixon, *No More Vietnams*, 146–47.
2. Conversation 700-002, 3 April 1972, 8:40–9:09 a.m., Oval Office.
3. Nixon, *No More Vietnams*, 147.
4. *Foreign Relations of the United States (FRUS), 1969–1976: Soviet Union, October 1971–May 1972*, ed. David C. Geyer, Nina D. Howland, and Kent Seig (Washington, DC: GPO, 2006), 14: Document 271 (hereafter *FRUS 1969–1976*, 14).
5. Ibid., Document 290.

"It Could Be a Bit Longer"

1. Kissinger, *White House Years*, 1304. Jussi Hanhimäki deserves credit for first uncovering most of the quotes below (Hanhimäki, "Some More 'Smoking Guns'? The Vietnam War and Kissinger's Summitry with Moscow and Beijing, 1971–1973," *Passport*, December 2001).
2. "Memorandum of Conversation," 20 June 1972, 2:05–6:05 p.m., "China—Dr. Kissinger's Visit, June 1972 Memcons" folder, NSCF Kissinger Office Files Box 97, RMNL.
3. "Memorandum of Conversation," 9 July 1971, 4:35–11:20 p.m., attached to Lord to Kissinger, "Memcon of Your Conversations with Chou En-lai," 29 July 1971, NSCF Box 1033, RMNL.
4. "Memorandum of Conversation," 20 June 1972, 2:05–6:05 p.m., "China—Dr. Kissinger's Visit, June 1972 Memcons" folder, NSCF Kissinger Office Files Box 97, RMNL.

5. "Memorandum of Conversation," 21 June 1972, 3:25–6:45 p.m., "China—Dr. Kissinger's Visit, June 1972 Memcons" folder, NSCF Kissinger Office Files Box 97, RMNL.

6. "Memorandum of Conversation," 22–23 June 1972, 11:03 p.m.–12:55 a.m., "China—Dr. Kissinger's Visit, June 1972 Memcons" folder, NSCF Kissinger Office Files Box 97, RMNL.

7. Zhou is quoted in Odd Arne Westad, Chen Jian, Stein Tonnesson, Nguyen Vu Tungand, and James Hershberg, eds., *77 Conversations between Chinese and Foreign Leaders on the Wars in Indochina, 1964–1977,* Working Paper No. 22, Cold War International History Project, May 1998, 179–82, http://wwics.si.edu/topics/pubs/ACFB39.pdf.

The Democrats

1. Theodore H. White, *The Making of the President 1972* (New York: Bantam, 1973), Kindle edition, chap. 7, "Confrontation at Miami."

2. Ibid., chap. 8, "The Eagleton Affair."

3. George McGovern, *Grassroots: The Autobiography of George McGovern* (New York: Random House, 1977), 193–94, 197–98, 223–24.

"No One Will Give a Damn"

1. Conversation 759-005, 2 August 1972, 10:34–11:47 a.m., Oval Office.

2. Ibid.

3. Kissinger, *Complete Memoirs E-book,* Kindle edition, chapter 31, "From Stalemate to Breakthrough: Testing the Stalemate."

4. Conversation 760-006, 3 August 1972, 8:28–8:57 a.m., Oval Office.

"Idealism with Integrity"

1. United Press International, "GOP Show Opens," *Los Angeles Times,* 21 August 1972, www.proquest.com.

2. Ronald Reagan speech, "Republican Convention," NBC, aired 21 August 1972, http://tvnews.vanderbilt.edu.

3. "Remarks on Accepting the Presidential Nomination of the Republican National Convention," 23 August 1972, *PPPUS: Nixon, 1972,* www.presidency.ucsb.edu/ws/?pid=3537.

4. Luu Van Loi and Nguyen Anh Vu, *Le Duc Tho–Kissinger Negotiations in Paris* (Hanoi: Gioi, 1996), 302–3. Randolph says the continued presence of Hanoi's troops in the South was another overriding goal (*Powerful and Brutal Weapons,* 325). Achievement of that goal had already been assured by the May 1971 proposal by Nixon and Kissinger for a ceasefire-in-place.

5. "Memorandum of Conversation," 26 September 1972, 10:30 a.m.–4:20 p.m., "Sensitive Camp David—Volume XVIII" folder, NSCF Box 856, RMNL. Kissinger simply dropped the "suitable interval" line from the American proposal the next day ("Memorandum of Conversation," 27 September 1972, 10:03 a.m.–3:38 p.m., "Sensitive Camp David—Volume XVIII" folder, NSCF Box 856, RMNL).

6. To say that Thieu thought he was winning the war was something of an

exaggeration. South Vietnamese leaders routinely credited battlefield successes to the strength and valor of their armed forces. At the same time, they lived in mortal fear of being forced to depend exclusively on those forces. Apart from that, Kissinger's summary of the basic conflict between Nixon's interests and Thieu's was accurate. It was the right time politically for Nixon to get out of Vietnam, but Saigon wanted him to stay in.

    7. Conversation 788-011, 29 September 1972, 3:16–3:30 p.m., Oval Office.

    8. Conversation 789-006, 30 September 1972, 10:56 a.m.–12:00 p.m., Oval Office.

    9. Conversation 790-008, 2 October 1972, 11:20–11:39 a.m., Oval Office.

    10. Haig to Kissinger, 4 October 1972, "Sensitive Camp David—Volume XIX" folder, NSCF Box 856, RMNL.

    11. "Memorandum of Conversation," 4 October 1972, 9:00 a.m.–12:50 p.m., "Sensitive Camp David—Volume XIX" folder, NSCF Box 856, RMNL. Haig's memoir account of this trip to Saigon is somewhat distorted. Nixon and Kissinger's proposals did not call for "the replacement of the existing government with a transitional body." The three-party commission was only supposed to work on arrangements for elections and was designed to fail, as we have seen, thanks to its requirement for unanimity (Haig, *Inner Circles,* 294–95).

"Our Terms Will Eventually Destroy Him"

    1. Conversation 793-006, 6 October 1972, 9:30–10:03 a.m., Oval Office.

    2. See "M'Govern Details a Foreign Policy Tied to 'Idealism,'" *New York Times,* 6 October 1972, www.proquest.com.

    3. See "The President's News Conference," 5 October 1972, *PPPUS: Nixon, 1972,* www.presidency.ucsb.edu/ws/?pid=3617.

    4. Murrey Marder, "Election, Peace Bid Separated," *Washington Post,* 6 October 1972, www.proquest.com.

    5. "One could imagine the public outcry if we rejected Hanoi's acceptance of our own proposals" (Kissinger, *White House Years,* 1348).

    6. Kissinger to Bunker, 5 October 1972, "Sensitive Camp David—Volume XIX" folder, NSC Box 856, RMNL.

    7. "Telegram from the Central Intelligence Agency Station in Saigon to the Agency," 5 October 1963, *FRUS, 1961–1963,* 4: Document 177 and Document 192. President John F. Kennedy personally oversaw the drafting of and approved the cable giving General Minh assurances that US aid would continue under the new South Vietnamese regime (Tape 114/A50, 8 October 1963, 5:30–6:15 p.m., Oval Office, John F. Kennedy Library, President's Office Files, Presidential Recordings Collection).

    8. Douglas Robinson, "Saigon Reports a Coup Attempt; Seizes Officers; But a U.S. Official Denies There Was Any Effort to Overthrow Thieu Regime," *New York Times,* www.proquest.com.

    9. See Situation Room to Haig, "Responses to Questions," 30 September 1972, "Sensitive Camp David—Volume XIX" folder, NSCF Box 856, RMNL.

    10. Kissinger to Bunker, "Deliver Immediately upon Receipt," 4 October 1972, "Sensitive Camp David—Volume XIX" folder, NSCF Box 856, RMNL.

11. Ibid.

12. Nixon, *No More Vietnams*, 82.

13. See "Address to the Nation on the Situation in Southeast Asia," 30 April 1970, *PPPUS: Nixon, 1970*, www.presidency.ucsb.edu/ws/?pid=2490.

14. Nixon, *No More Vietnams*, 136.

15. Conversation 793-006, 6 October 1972, 9:30–10:03 a.m., Oval Office.

16. Richard Nixon, *Leaders* (New York: Warner, 1982), 42, 46, 47, 57, 58, 59, 65, 66, 84.

17. Ibid., 76; Robert Donovan, "Nixon, De Gaulle Trade Warm Greetings Despite Differences," *New York Times*, 1 March 1969, www.proquest.com.

18. *Haldeman Diaries*, 28 February 1969.

19. Nixon, *Leaders*, 48, 76, 78–79; Nixon, *RN*, 374.

20. "Memorandum of Conversation," 22 February 1972, 2:10–6:00 p.m., "Beginning February 20, 1972" folder, President's Office Files, Box 87, RMNL.

21. Nixon, *Leaders*, 61; Henry Tanner, "De Gaulle to Visit the U.S. in '70," *New York Times*, 3 March 1969, www.proquest.com.

22. Conversation 793-006, 6 October 1972, 9:30–10:03 a.m., Oval Office.

23. Ibid.

Blowup 1968

1. See Hughes, *Chasing Shadows*, 1–56.

"We're behind the Trees!"

1. See "Memorandum of Conversation," 8 October 1972, 10:30 a.m.–7:38 p.m., "Sensitive—Vol. XX, C.D. Memcons, Oct. 8–11 & Oct. 17, 1972" folder, NSCF Box 856, RMNL. Kissinger, *White House Years*, 1335, 1341–50.

2. Kissinger, *Ending the Vietnam War*, 550.

3. Seymour Hersh, *The Price of Power* (New York: Summit Books, 1983), 584.

4. Ibid., 584–85.

5. Walter Isaacson, *Kissinger: A Biography* (New York: Simon and Schuster, 1992), 449.

6. See "Memorandum of Conversation," 9 October 1972, 3:58–6:08 p.m., "Sensitive—Vol. XX, C.D. Memcons, Oct. 8–11 & Oct. 17, 1972" folder, NSCF Box 856, RMNL.

"Saving Face or Saving Lives"

1. Ernest R. May and Janet Fraser, eds., *Campaign '72: The Managers Speak* (Cambridge: Harvard University Press, 1973), 229.

2. "Transcript of Senator McGovern's Speech Offering a Plan for Peace in Indochina," *New York Times*, 11 October 1972; Jules Witcover, "McGovern Outlines Plan to End War," *Los Angeles Times*, 11 October 1972, www.proquest.com.

3. Conversation 364-004, 5 October 1972, 12:26–1:36 p.m., Executive Office Building.

"Brutalize Him"

1. See H. R. Haldeman and Joseph DiMona, *The Ends of Power* (New York: Times Books, 1978), 75.

2. *Haldeman Diaries*, 12 October 1972; Nixon, *RN*, 691.

3. Nixon wrote that "Haig seemed rather subdued" in *RN*, 693. "A better word might have been *despondent*," wrote Haig in *Inner Circles*, 300.

4. See Kissinger, *White House Years*, 1347.

5. Associated Press, "French Mission Hit in U.S. Hanoi Raid; 6 Killed, Chief Diplomat Hurt; Paris Protests," *Los Angeles Times*, 12 October 1972; Bernard Gwertzman, "U.S. Is Regretful: But Pentagon Says a Hanoi Missile May Have Caused Blast," *New York Times*, 12 October 1972, www.proquest.com.

6. Nixon danced delicately around the word "reparations" in his memoir account of this meeting: "The Communists tried to claim that this money would be reparations for the war they charged we had unleashed upon them; but however they tried to justify it, taking money from the United States represented a collapse of communist principle" (Nixon, *RN*, 692).

7. "Memorandum of Conversation," 26 September 1972, 10:30 a.m.–4:20 p.m., "Sensitive Camp David—Volume XVIII" folder, NSCF Box 856, RMNL.

8. Conversation 366-006, 12 October 1972, 6:10–8:46 p.m., Executive Office Building.

9. *Haldeman Diaries*, 12 October 1972.

10. See Haig, *Inner Circles*, 299.

11. *Haldeman Diaries*, 12 October 1972. See Haig, *Inner Circles*, 300.

12. *Washington Star-News*, "Thieu Bars Coalition and Would Kill Foe 'to Last Man,'" *New York Times*, 13 October 1972, www.proquest.com.

13. Joseph Kraft, "The Vietnam Timetable," 8 October 1972, *Washington Post*, www.proquest.com.

14. "Max Frankel/Mr. Kissinger," 13 October 1972, 9:16 p.m., Kissinger Telcons Box 16, RMNL.

15. "Bob Toth/Mr. Kissinger," 13 October 1972, 4:20 p.m., Kissinger Telcons Box 16, RMNL.

16. Conversation 149-14, 15 October 1972, 12:00–12:14 p.m., Camp David Study Table.

17. Conversation 798-004, 14 October 1972, 10:03–10:39 a.m., Oval Office.

18. Carl Bernstein and Bob Woodward, "FBI Finds Nixon Aides Sabotaged Democrats," *Washington Post*, 10 October 1972, www.proquest.com.

19. Bob Woodward and Carl Bernstein, "Lawyer for Nixon Said to Have Used GOP's Spy Fund," *Washington Post*, 16 October 1972, www.proquest.com.

20. Carl Bernstein and Bob Woodward, "Key Nixon Aide Named as 'Sabotage' Contact," *Washington Post*, 15 October 1972, www.proquest.com.

21. *Haldeman Diaries*, 16 October 1972.

22. Conversation 799-009, 16 October 1972, 9:33–10:12 a.m., Oval Office.

23. John Chancellor and John Cochran, "Campaign '72 / Nixon on POW's and Amnesty," *NBC Evening News*, aired 16 October 1972, http://tvnews.vanderbilt.edu; "Remarks to a Meeting of the National League of Families of American Prisoners and Missing in Southeast Asia," 16 October 1972, *PPPUS: Nixon, 1972*, www.presidency.ucsb.edu/ws/?pid=3632.

24. James M. Naughton, "McGovern Sees 'Sabotage,'" *New York Times*, 17 October 1972, www.proquest.com.

25. Conversation 031-084, 16 October 1972, 7:20–7:21 p.m., White House Telephone.

Kissinger v. Thieu

1. Kissinger, *White House Years,* 1364.
2. Kissinger to Haig, 17 October 1972, 8:48 p.m., "HAK's Saigon Trip Oct 16–Oct 23, 1972 HAKTO & TOHAK Cables [2 of 2]" folder, NSCF Kissinger Office Files Box 104, RMNL.
3. "Memorandum of Conversation," 17 October 1972, 10:37 a.m.–10:10 p.m., "Camp David Vol. XX," NSCF Box 856, Digital National Security Archive Item Number KT00584.
4. Kissinger, *White House Years,* 1364–65.
5. See Isaacson, *Kissinger,* 363.
6. Kissinger, *White House Years,* 1369.
7. Ibid., 1320.
8. "Memorandum of Conversation," 19 October 1972, 9:10–12:20 p.m., "Sensitive Camp David—Volume XX" folder, NSCF Box 857, RMNL.
9. Isaacson, *Kissinger,* 463.
10. Haig to Nixon, "Meeting with President Thieu," 19 October 1972, "HAK Paris/Saigon Trip 16–23 Oct 1972" folder, NSCF Box 25, RMNL; "Memorandum of Conversation," 19 October 1972, 9:10–12:20 p.m., "Sensitive Camp David—Volume XX" folder, NSCF Box 857, RMNL.
11. Nixon, *RN,* 697; Kissinger, *White House Years,* 1371.
12. Guay to Haig, 19 October 1972, 5:14 p.m., "Sensitive Camp David—Volume XX" folder, NSCF Box 857, RMNL.
13. Kissinger, *White House Years,* 1371.
14. Haig to Nixon, "Message from the North Vietnamese," 20 October 1972, "Sensitive Camp David—Volume XX" folder, NSCF Box 857, RMNL.
15. Haig to Nixon, "Meeting with President Thieu," 19 October 1972, "HAK Paris/Saigon Trip 16–23 Oct 1972" folder, NSCF Box 25, RMNL.
16. "Memorandum of Conversation," 20 October 1972, 2:10–5:35 p.m., "Sensitive Camp David—Volume XX" folder, NSCF Box 857, RMNL.
17. Conversation 799-023, 16 October 1972, 3:59–4:30 p.m., Oval Office.
18. "Third Annual Report to the Congress on United States Foreign Policy," 9 February 1972, *PPPUS: Nixon, 1971,* www.presidency.ucsb.edu/ws/?pid=3736.
19. Conversation 823-001, 14 December 1972, 9:59–11:46 a.m., Oval Office; Thieu to Nixon, 26 November 1972, "Sensitive C.D.—Volume XXII (1)" folder, NSCF Box 858, RMNL.
20. See Conversation 799-023, 16 October 1972, 3:59–4:30 p.m., Oval Office.
21. "Memorandum of Conversation," 20 October 1972, 2:10–5:35 p.m., "Sensitive Camp David—Volume XX" folder, NSCF Box 857, RMNL.
22. Kissinger to White House, attached to Haig to Nixon, 20 October 1972,
23. Kissinger, *White House Years,* 1377.
24. Ibid., 1377–78.

"No Possibility Whatever"

1. Memorandum of Conversation, 21 October 1972, 10:16 a.m.–1:10 p.m., "Sensitive Camp David—Volume XX" folder, NSCF Box 857, RMNL.
2. Conversation 806-002, 23-24 October 1972, 11:35 p.m.–12:05 a.m., Oval Office; Kissinger, *White House Years,* 1379.
3. Kissinger, *White House Years,* 1380.

"The Man Who Should Cry Is I"

1. Bunker to Haig, 22 October 1972, Digital National Security Archive Item Number KT00589; Kissinger, *White House Years,* 1382.
2. Kissinger, *White House Years,* 1382.
3. Kissinger to Haig, 22 October 1972, "NCS—Top Secret" folder, WHSF-SMOF Haldeman Box 180, RMNL, www.nixonlibrary.gov/virtuallibrary/releases/dec10/29.pdf and Digital National Security Archive Item Number KT00590; Kissinger, *White House Years,* 1385.
4. See Nixon, *RN,* 702.
5. *Haldeman Diaries,* 22 October 1972.
6. *Newsweek* excerpt, "The Aim: 'A 3-Sided Coalition,'" *Washington Post,* 22 October 1972, www.proquest.com; Kissinger, *White House Years,* 1388–89; Kissinger to Haig, 22 October 1972, "HAK Paris/Saigon Trip; HAKTO 16–23 Oct 72" folder, NSCF Kissinger Office Files Box 25, RMNL.
7. Nguyen, *Hanoi's War,* Kindle edition, chap. 8, "War for Peace: Sabotaging Peace."
8. "Dr. Kissinger's Meeting with President Thieu, October 23, 1972," attached to Haig to Nixon, "Meeting with President Thieu," 23 October 1972, "Sensitive Camp David—Volume XX" folder, NSCF Box 857, RMNL.
9. "Secretary Peterson/Kissinger," 25 October 1972, 11:00 a.m., Kissinger Telcons Box 16, RMNL.

"The Fellow Is Off His Head"

1. "Ambassador Dobrynin/Mr. Kissinger," 23 October 1972, 11:22 p.m.; and "Ambassador Dobrynin/HAK," 15 October 1972, 8:35 p.m., both in Kissinger Telcons Box 16, RMNL.
2. Conversation 806-001, 23 October 1972, 11:20–11:35 p.m., Oval Office.
3. Conversation 371-019, 23 October 1972, 8:34–9:17 a.m., Executive Office Building. Credit goes to Jeffrey Kimball for discovering this passage.
4. Louis Harris, "Nixon Has Lead on Issues," *Chicago Tribune,* 17 July 1972, www.proquest.com.
5. Conversation 806-001, 23 October 1972, 11:20–11:35 p.m., Oval Office.
6. Negroponte to Kissinger, 23 October 1972, "Sensitive; Camp David—Volume XXI" folder, NSCF Box 857, RMNL.
7. Conversation 806-002, 23-24 October 1972, 11:35 p.m.–12:05 a.m., Oval Office.
8. "Veterans Day Message," 23 October 1972, *PPPUS: Nixon, 1972,* www.presidency.ucsb.edu/ws/?pid=3644.
9. Conversation 801-004, 17 October 1972, 8:36–9:59 a.m., Oval Office.

10. Conversation 806-002, 23–24 October 1972, 11:35 p.m.–12:05 a.m., Oval Office.

11. "Ambassador Dobrynin/Mr. Kissinger," 24 October 1972, 12:10 a.m., Kissinger Telcons Box 16, RMNL.

12. "Memorandum of Conversation," 24 October 1972, 6:55–7:45 p.m., Digital National Security Archive Item Number KT00592.

13. James F. Clarity, "Nixon Honors Veterans with a Fellowship," *New York Times*, 25 October 1972, www.proquest.com; Carroll Kilpatrick, "Nixon Vows Jobs for Vietnam Veterans," *Washington Post*, 25 October 1972, www.proquest.com. The *Toledo Blade* ran the AP story under the enthusiastic, but ultimately false, headline: "Nixon Puts Action Where Signature Is; Army Veteran Gets White House Fellowship on the Spot," 25 October 1972, http://news.google.com/newspapers?nid=1350&dat=19721025&id=6uhOAAAAIBAJ&sjid=9QEEAAAAIBAJ&pg=6598,2076598.

14. "18 White House Fellows Selected; Enthusiasm Undimmed by Watergate," *New York Times*, 22 May 1973, www.proquest.com.

15. Nixon, *RN*, 704.

No Coalition Government

1. Tom Streithorst, "Vietnam/Thieu and Peace; Kissinger," *NBC Evening News*, aired 24 October 1972, http://tvnews.vanderbilt.edu. R. W. Apple Jr., "Nixon Would Bar Forced Coalition in South Vietnam," *New York Times*, 28 October 1968, www.nytimes.com.

"Coalition is a code word in international settlements, and wherever there have been coalition governments that include Communists it usually means that the Communists have, of course, prevailed and eventually expelled, if I may use that term too, expelled the non-Communists from the government" ("The President's News Conference," 20 July 1970, *PPPUS: Nixon, 1970*, www.presidency.ucsb.edu/ws/?pid=2588).

"We will not negotiate with the enemy for accomplishing what they cannot accomplish themselves and that is to impose against their will on the people of South Vietnam a coalition government with the Communists" ("The President's News Conference," 29 June 1972, *PPPUS: Nixon, 1972*, www.presidency.ucsb.edu/ws/?pid=3480).

2. Craig R. Whitney, "Speech in Saigon: Ceasefire Obstacles Seen, but President Expects Agreement," *New York Times*, 25 October 1972, www.proquest.com.

3. Jules Witcover, "Could Have Had Peace in 1968, McGovern Says; Democrat Charges Nixon with Continuing War to Avoid Hawks' Criticism," *Washington Post*, 25 October 1972, www.proquest.com; George Lardner Jr., "McGovern: Viet Peace Could 'Destroy' Nixon," *Washington Post*, 25 October 1972, www.proquest.com.

4. "Remarks on Election Eve," 6 November 1972, *PPPUS: Nixon, 1972*, www.presidency.ucsb.edu/ws/?pid=3701. He made nearly identical remarks in seven campaign speeches delivered on 2–4 November 1972.

5. Carl Bernstein and Bob Woodward, "Testimony Ties Top Nixon Aide to Secret Fund," *Washington Post*, 25 October 1972, www.proquest.com.

6. Carl Bernstein and Bob Woodward, *All the President's Men* (New York: Simon and Schuster, 1974), Kindle edition, chap. 9.

7. Vanderbilt Television News Archive, 25 October 1972, http://openweb.tv news.vanderbilt.edu.

"Peace Is at Hand"

1. Conversation 807-007, 26 October 1972, 9:22–9:54 a.m., Oval Office.
2. See "CBS News Special: Kissinger News Briefing on the War," aired 26 October 1972, Reference Cassette WHCA #5843; and "Transcript of Kissinger's News Conference on the Status of the Ceasefire Talks," *New York Times*, 27 October 1972, www.nytimes.com.
3. See John J. O'Conner, "CBS Wins Network Race on Kissinger Briefing," *New York Times*, 27 October 1972, www.nytimes.com.
4. Isaacson, *Kissinger*, 459.
5. *Haldeman Diaries*, 26 October 1972.
6. See Conversation 032-063, 26 October 1972, 11:44–11:53 p.m., White House Telephone.
7. Conversation 032-065, 26–27 October 1972, 11:53 p.m.–12:21 a.m., White House Telephone.
8. Jules Witcover, "McGovern Sees Peace Terms as Similar to France's in 1954," *Los Angeles Times*, 27 October 1972, www.proquest.com.
9. Jack Perkins, "Campaign '72/McGovern on Vietnam Peace," *NBC Evening News*, aired 26 October 1972, http://tvnews.vanderbilt.edu.
10. Jules Witcover, "McGovern Sees Peace Terms as Similar to France's in 1954," *Los Angeles Times*, 27 October 1972, www.proquest.com.
11. Douglas E. Kneeland, "McGovern Says He Hopes Nixon View Is Confirmed," *New York Times*, 27 October 1972, www.proquest.com.
12. Conversation 032-065, 26–27 October 1972, 11:53 p.m. –12:21 a.m., White House Telephone.
13. "Senator James Buckley/Mr. Kissinger," 26 October 1972, 6:25 p.m., Kissinger Telcons Box 16, RMNL.
14. "Amb. Bush/Kissinger," 28 October 1972, 1:35 p.m., Kissinger Telcons Box 16, RMNL.
15. "Gov. Ronald Reagan/Mr. Kissinger," 27 October 1972, 11:55 a.m., Kissinger Telcons Box 16, RMNL.
16. "Senator Mansfield/Kissinger," 28 October 1972, 2:00 p.m., Kissinger Telcons Box 16, RMNL.
17. "McG. Bundy/Mr. Kissinger," 26 October 1972, 2:34 p.m., Kissinger Telcons Box 16, RMNL.
18. Associated Press, "Nixon Political Gain Seen by Humphrey," *Los Angeles Times*, 27 October 1972, www.proquest.com.
19. "James Reston/Mr. Kissinger," 26 October 1972, 3:05 p.m., Kissinger Telcons Box 16, RMNL.
20. Associated Press, "Evolving Peace Terms Resemble '69 Kissinger Plan," *New York Times*, 27 October 1972, www.proquest.com.
21. Henry A. Kissinger, "The Viet Nam Negotiations," *Foreign Affairs* 47, no. 2 (January 1969): 211–34, Academic Search Complete, EBSCOhost.
22. Associated Press, "Evolving Peace Terms Resemble '69 Kissinger Plan," *New York Times*, 27 October 1972, www.proquest.com.

"A Little Bit Diabolically"

1. United Press International, "'There'll be no S. Vietnam Peace until I Sign'—Thieu," *Los Angeles Times*, 27 October 1972, www.proquest.com.
2. Craig R. Whitney, "Doubt Is Voiced; South Vietnam Resists Accord against Its People's Interests," *New York Times*, 27 October 1972, www.proquest.com.
3. "Thieu Insists That a Cease-Fire Hinges on Hanoi Troop Pullout," *New York Times*, 28 October 1972, www.proquest.com.
4. Conversation 810-001, 31 October 1972, 11:04–11:10 a.m., Oval Office.
5. Conversation 810-002, 31 October 1972, 11:10–11:36 a.m., Oval Office.
6. "Peace at Last?," *New York Times*, 27 October 1972, www.proquest.com.
7. Richard L. Madden, "Rockefeller Stumps for Nixon; Hints He Will Seek 5th Term," *New York Times*, 1 November 1972, www.proquest.com.
8. See Conversation 810-005, 31 October 1972, 2:52–3:23 p.m., Oval Office.
9. Craig Whitney, "Thieu Calls Draft Accord 'Surrender to the Communists,'" *New York Times*, 1 November 1972, www.proquest.com.
10. Vanderbilt Television News Archive, 1 November 1972, http://tvnews.vanderbilt.edu.
11. Christopher Lydon, "Shriver Criticizes Proposal for Coalition in Vietnam," *New York Times*, 1 November 1972; Stuart Auerbach, "Shriver Hits Viet Plan as a Sellout by Nixon," *Washington Post*, 1 November 1972, www.proquest.com.
12. See Christopher Lydon, "Shriver Dubious about War Pact," *New York Times*, 2 November 1972, www.proquest.com.
13. Richard M. Cohen, "Shriver, in Baltimore, Assails Nixon," *Washington Post*, 3 November 1972, www.proquest.com.
14. "The Junior Partners," *Time*, 6 November 1972, 54, *Academic Search Complete*, EBSCOhost.
15. "He accused the South Vietnamese president, Nguyen Van Thieu, of jeopardizing peace negotiations" (Frank Lynn, "M'Govern Draws a Cheering 20,000 in Garment Area," *New York Times*, 2 November 1972, www.proquest.com).
16. See Nicholas C. Chriss, "McGovern Calls Peace Move 'Cruel Deception,'" *Los Angeles Times*, 4 November 1972; and James M. Naughton, "M'Govern Asserts Nixon Pretended to Be Near Peace," *New York Times*, 4 November 1972, www.proquest.com.
17. Christopher Lydon, "Shriver Criticizes Proposal for Coalition in Vietnam," *New York Times*, 1 November 1972, www.proquest.com.
18. See Christopher Lydon, "Shriver Dubious about War Pact," *New York Times*, 2 November 1972, www.proquest.com.
19. Craig Whitney, "Thieu Calls Draft Accord 'Surrender to the Communists,'" *New York Times*, 1 November 1972, www.proquest.com.
20. "Peace Delays Help No One," *Los Angeles Times*, 3 November 1972, www.proquest.com.

21. Donald Kirk, "In-Fighter Thieu: Champ Is Unpopular," *Chicago Tribune*, 6 November 1972, www.proquest.com.

22. Robert Keatley, "U.S. Says Vietnam Accord Isn't Imperiled Despite Sparring," *Wall Street Journal*, 3 November 1972, www.proquest.com.

23. Victor Zorza, "The Real Issues of Renegotiation," *Washington Post*, 1 November 1972, www.proquest.com.

24. Henry Kamm, "Gen. Minh Opposes Draft Agreement," *New York Times*, 2 November 1972, www.proquest.com.

25. See Thomas Lippman, "Thieu Calls Draft Pact a Sellout," *Washington Post*, 1 November 1972, www.proquest.com.

26. Henry Kamm, "Gen. Minh Opposes Draft Agreement," *New York Times*, 2 November 1972, www.proquest.com.

The Chennault Affair

1. See Robert B. Semple Jr., "Nixon Denounces Welfare Inequity, Calls for National Standards—Repudiates Criticism of Johnson Peace Efforts," *New York Times*, 26 October 1968; and Peter H. Silberman, "Nixon Reports Cease-Fire Hint," *Washington Post*, 26 October 1968, www.proquest.com.

"The Clearest Choice"

1. Douglas E. Kneeland, "M'Govern Warns Nixon Lacks Plan to Quit Vietnam," *New York Times*, 5 November 1972, www.proquest.com.

2. So successful was Nixon at concealing extreme partisanship behind a nonpartisan pose that one scholar who has devoted a long and celebrated career to the study of political rhetoric wrote that "Nixon's election eve speech is an almost nonpartisan get-out-the-vote appeal that specifies the issues he hopes voters will keep in mind as they ballot" and "an almost nonpartisan appeal to the country to vote" (Kathleen Hall Jamieson, *Packaging the Presidency: A History and Criticism of Presidential Campaign Advertising* [New York: Oxford University Press, 1996], Kindle edition, chap. 6, "1968: The Competing Pasts of Nixon and Humphrey").

3. "Remarks on Election Eve," 6 November 1972, *PPPUS: Nixon, 1972*, www.presidency.ucsb.edu/ws/?pid=3701.

Election Day 1972

1. Nixon, *RN*, 717.

2. Ibid., 714.

3. Theodore H. White, *The Making of the President 1972* (New York: Bantam, 1973), Kindle edition, chap. 1, "The Solitary Man."

4. Ibid.

5. Nixon, *RN*, 715, 717.

6. Ibid., 715.

7. White, *The Making of the President 1972*, Kindle edition, chap. 13, "Appeal to the People: Verdict in November."

8. Grantland Rice, "Alumnus Football," *Pittsburgh Press*, 2 November 1914, http://news.google.com/newspapers.

9. "Remarks on Being Reelected to the Presidency," 7 November 1972, *PPPUS: Nixon, 1972*, www.presidency.ucsb.edu/ws/?pid=3702, YouTube, http://youtube/8PimoaXdTsE?list=PL23130A5F7BDC2E09.

10. "Transcript of the Speech by McGovern," *New York Times*, 8 November 1972, www.proquest.com.

11. White, *The Making of the President 1972*, Kindle edition, chap. 13, "Appeal to the People: Verdict in November."

12. Conversation 033-060, 8 November 1972, 1:16–1:28 a.m., White House Telephone.

13. Ibid.

Promises and Threats

1. Nixon to Thieu, 8 November 1972, "Sensitive Camp David—Volume XXI (1)" folder, NSCF Box 857, RMNL.

2. Ibid.

3. "Memorandum of Conversation," 20 November 1972, 10:45 a.m.–4:55 p.m., "C.D. Vol. XXI—Minutes of Meetings, Paris, Nov. 20–Nov. 25, 1972" folder, NSCF Box 858, RMNL.

4. "Memorandum of Conversation," 21 November 1972, 3:02–7:26 p.m., "C.D. Vol. XXI—Minutes of Meetings, Paris, Nov. 20–Nov. 25, 1972" folder, NSCF Box 858, RMNL.

5. "Memorandum of Conversation," 24 November 1972, 11:00 a.m.–12:20 p.m., "C.D. Vol. XXI—Minutes of Meetings, Paris, Nov. 20–Nov. 25, 1972" folder, NSCF Box 858, RMNL.

6. Kissinger, *White House Years*, 1419.

7. Nick Thimmesch, "Dr. Henry Kissinger Gives an Interview," *Chicago Tribune*, 26 November 1972, www.proquest.com.

8. See Conversation 816-003, 29 November 1972, 2:52–5:32 p.m., Oval Office; "Memorandum of Conversation," 29 November 1972, 3:05–5:10 p.m., "GVN Memcons, Nov. 20, 1972–Apr. 3, 1973 [2 of 3]" folder, NSCF Kissinger Office Files Box 104, RMNL. Nixon mentioned that the threat of a cutoff came from the political *right* in *RN*, 724, but Kissinger leaves out that crucial point in *White House Years*, 1425–26. See also Nixon to Thieu, 24 November 1972, in "Memorandum of Conversation," 24 November 1972, 7:30–8:45 p.m., "C.D. Vol. XXI—Minutes of Meetings, Paris, Nov. 20–Nov. 25, 1972" folder, NSCF Box 858, RMNL.

9. Conversation WH6811-04-13723-13724-13725, 8 November 1968, 9:23 p.m., Mansion.

10. Conversation 817-016, 30 November 1972, 12:17–1:11 p.m., Oval Office.

Christmas Bombing

1. See Nixon, *No More Vietnams*, 157.

2. Conversation 806-001, 23 October 1972, 11:20–11:35 p.m., Oval Office.

3. See Conversation 810-005, 31 October 1971, 2:52–3:23 p.m., Oval Office.

4. "The President," 6 November 1972, 10:00 a.m., Kissinger Telcons Box 17, RMNL.

5. See "Mr. Kissinger/Governor Reagan," 15 December 1972, 11:00 a.m., Kissinger Telcons Box 17, DNSA http://gateway.proquest.com/openurl?url_ver=Z39.88-2004&res_dat=xri:dnsa&rft_dat=xri:dnsa:article:CKA09150. Seth Center quoted this transcript in "Confronting Decline: The Resilience of the U.S. Conception of America's Role in the World, 1968–1975" (PhD diss., University of Virginia, 2011), Proquest UMI 3484467.

6. Kissinger explained Le Duc Tho's position to the Chinese: "He has demanded that we return to the old agreement without change or to a new agreement in which he proposes so many significant changes that it will be worse than the old one." The North could play the carrot-and-stick game too ("Memorandum of Conversation," 7–8 December 1972, 11:25 p.m.–12:15 a.m., Digital National Security Archive Item Number KT00632).

7. "Memorandum of Conversation," 7 December 1972, 3:00–7:00 p.m., Digital National Security Archive Item Number KT00630; Kimball, *Nixon's Vietnam War*, 358.

8. "Memorandum of Conversation," 12 December 1972, 3:07–7:35 p.m., Digital National Security Archive Item Number KT00640.

9. "Memorandum of Conversation," 13 December 1972, 10:30 a.m.–4:24 p.m., Digital National Security Archive Item Number KT00642.

10. Conversation 034-069, 13 December 1972, 8:55–9:07 p.m., White House Telephone.

11. See Haig to Kissinger, 13 December 1972, "Sensitive C.D.—Volume XXII (2)" folder, NSCF Box 858, RMNL, www.scribd.com/doc/16332927/Dec-13-1972-Haig-to-Kissinger.

12. Nixon, *RN*, 735.

13. Webb and Poole, *The Joint Chiefs of Staff and the War in Vietnam, 1971–1973*, 293–99.

14. Stanley Karnow, *Vietnam: A History* (New York: Penguin, 1991), 667–68.

15. Webb and Poole, *The Joint Chiefs of Staff and the War in Vietnam, 1971–1973*, 300.

16. Kissinger, *White House Years*, 1371.

17. Kissinger, *Complete Memoirs E-book*, Kindle edition, chap. 33, "Peace Is at Hand: The January Round"; Kissinger, *Ending the Vietnam War*, 424–25.

18. Transcript of Kissinger's News Conference on the Status of the Ceasefire Talks," *New York Times*, 27 October 1972, www.nytimes.com.

19. Conversation 793-006, 6 October 1972, 9:30–10:03 a.m., Oval Office.

20. Kissinger, *Ending the Vietnam War*, 425.

21. Conversation 366-006, 12 October 1972, 6:10–8:46 p.m., Executive Office Building.

22. Kissinger, *Ending the Vietnam War*, 425.

23. Webb and Poole, *The Joint Chiefs of Staff and the War in Vietnam, 1971–1973*, 300.

24. "House Democrats Condemn Nixon's Policies on Vietnam War in a 154–75 Caucus Vote," *Wall Street Journal*, 3 January 1973, www.proquest.com.

25. Spencer Rich, "Democrats Vote to Bar War Funds," *Washington Post*, 5 January 1973, www.proquest.com.

26. See Conversation 823-001, 14 December 1972, 9:59–11:46 a.m., Oval Office.

27. Richard L. Lyons, "Senate Unit Set to Act on Jan. 20," *Washington Post*, 3 January 1973, www.proquest.com.

28. Spencer Rich, "Democrats Vote to Bar War Funds," *Washington Post*, 5 January 1973, www.proquest.com.

29. Conversation 816-003, 29 November 1972, 2:52–5:32 p.m., Oval Office; "Memorandum of Conversation," 29 November 1972, 3:05–5:10 p.m., "GVN Memcons, Nov. 20, 1972–Apr. 3, 1973 [2 of 3]" folder, NSC Kissinger Office Files Box 104, RMNL.

30. "Mr. Kissinger/The President," 18 January 1973, 9:40 a.m., Digital National Security Archive Item Number KA09303.

31. "Senator Goldwater/Mr. Kissinger," 18 January 1973, 11:13 a.m., Digital National Security Archive Item Number KA09307.

32. Ibid. In his 1979 memoir, Kissinger wrote that Goldwater and Stennis acted at "our" suggestion, but in 2003's *Ending the Vietnam War*, he changed "our" to "Nixon's" in a sentence otherwise identical to the one in the earlier book (see Kissinger, *White House Years*, 1470; and Kissinger, *Ending the Vietnam War*, 428). He was once again trying to hide his own hand, but not Nixon's.

33. David E. Rosenbaum, "Goldwater and Stennis Tell Saigon Not to Balk," *New York Times*, 19 January 1973, www.nytimes.com.

34. Kissinger, *Ending the Vietnam War*, 424.

35. Program listings for the 18 January 1973 newscasts on CBS, NBC, and ABC, Vanderbilt Television News Archive, http://tvnews.vanderbilt.edu.

36. See Conversation 036-021, 20 January 1973, 9:32–9:59 a.m., White House Telephone.

37. "Oath of Office and Second Inaugural Address," 20 January 1973, *Public Papers of the Presidents of the United States: Richard M. Nixon, 1973* (Washington, DC: GPO, 1975), www.presidency.ucsb.edu/ws/?pid=4141 (hereafter *PPPUS: Nixon, 1973*).

38. See Kissinger to Bunker, 20 January 1973, *FRUS 1969–1976: Vietnam, October 1972–January 1973*, 9: Document 313.

39. See Nguyen Tien Hung and Jerrold L. Schecter, *The Palace File* (New York: Harper and Row, 1986), 155.

40. Ibid.

"Let Us Be Proud"

1. "Address to the Nation Announcing Conclusion of an Agreement on Ending the War and Restoring Peace in Vietnam," 23 January 1973, *PPUS: Nixon, 1973*, www.presidency.ucsb.edu/ws/?pid=3808.

2. George Gallup, "Cease-Fire Gives Nixon 68% Again," *Washington Post*, 6 February 1973, www.proquest.com.

3. George Gallup, "40% Favor Rebuilding N. Vietnam," *Washington Post*, 26 January 1973, www.proquest.com. Even then, after the bombing but before announcement of the accords, Nixon's approval rating was 54 percent.

4. George Gallup, "Public Hails Peace Accord, Opposes a New Involvement," *Washington Post*, 30 January 1973, www.proquest.com.

5. The impression persists that the Christmas bombing had more than a temporary impact on public approval of Nixon, despite the poll results showing very high ratings for him (and the bombing) once the Paris Accords were announced in January 1973. Nixon did enjoy a few months of great popularity after the Paris Accords announcement and before the congressional Watergate investigations became daily, televised events. Some writers, however, portray a steep, uninterrupted decline: "Critics dubbed the December air attacks the Christmas bombings, and Democratic congressional leaders promised to take immediate action to prevent a continuation of the American military role in Vietnam. Nixon, after winning a landslide victory in November, saw his approval ratings plummet into the mid-30 percent range, a level they never recovered from with the revelations about Watergate. Thus, for Nixon to achieve his goal of a settlement that kept Thieu in power and allowed for future American military strikes, he needed to force Saigon to comply with the treaty negotiated by the United States" (David F. Schmitz, *Richard Nixon and the Vietnam War: The End of the American Century* [Lanham, MD: Rowman and Littlefield, 2014], Adobe Digital Edition, chap. 6, "Denouement"). Nixon's 68 percent approval rating of 26–29 January 1973 and 65 percent approval rating of 16–19 February 1973 reflect public opinion after the Christmas bombing and Paris Peace Accords announcement and before the long Watergate decline (George Gallup, "Nixon Popularity Remains Near Viet Cease-Fire Peak," *Washington Post,* 10 March 1973, www.proquest.com).

6. Sylvan Fox, "Fighting Goes On; Saigon Reports Slight Decline in Level of Military Activity," 1 February 1973, *New York Times,* www.proquest.com.

7. Sylvan Fox, "Cease-Fire Violations Said to Be Minor: Toll of 4,295 Reported," *New York Times,* www.proquest.com.

8. Henry Kamm, "Cease-Fire and 10,000 Corpses," *New York Times,* 25 February 1973, www.proquest.com.

9. Conversation 879-005, 14 March 1973, 11:10 a.m.–12:25 p.m., Oval Office.

10. Conversation 881-002, 16 March 1973, 10:18–10:33 a.m., Oval Office.

The Prisoners Dilemma

1. Nixon to Thieu, 29 October 1972, 14 November 1972, and 5 January 1972, qtd. in Hung and Schecter, *The Palace File,* 380, 386, and 392. Hung and Schecter helpfully provide facsimiles of these letters and others between Nixon and Thieu.

2. Kennedy to Kissinger, "Thieu's Comments on Agreement," 20 December 1972, Digital National Security Archive, Item Number VW01169.

3. Webb and Poole, *The Joint Chiefs of Staff and the War in Vietnam, 1971–1973,* 300.

4. Conversation 876-013/877-001, 12 March 1973, 4:24–6:34 p.m., Oval Office.

5. Nixon actually *wanted* to be viewed as someone who would use excessive force against North Vietnam. "I call it the Madman Theory," he told Haldeman. "I want the North Vietnamese to believe I've reached the point where I might do *anything* to stop the war. We'll just slip the word to them that, 'for God's sake, you know Nixon is obsessed about Communism. We can't restrain him when

he's angry—and he has his hand on the nuclear button'—and Ho Chi Minh himself will be in Paris in two days begging for peace."[12] At least with American liberals, Nixon did succeed in crafting a Madman image (H. R. Haldeman and Joseph DiMona, *The Ends of Power* [New York: Times Books, 1978], 82–83). For a complete discussion of Madman Theory, including the views of those who doubt Nixon practiced it, see Jeffrey Kimball, *The Vietnam War Files* (Lawrence: University Press of Kansas, 2004), 15–20. His tapes leave no doubt: "Henry's got them convinced there's a madman in the White House who doesn't care about polls or anything and might bomb you again" (Conversation 846-004, 1 February 1973, 10:35 a.m.–12:35 p.m., Oval Office).

6. Conversation 846-004, 1 February 1973, 10:35 a.m.–12:35 p.m., Oval Office.

The Final Cutoff

1. "Veto of the Supplemental Appropriations Bill Containing a Restriction on United States Air Operations in Cambodia," 27 June 1973, *PPPUS: Nixon, 1973*, www.presidency.ucsb.edu/ws/?pid=3883.

2. John Herbers, "Test Intensifies; Rider Put on Another Bill after the House Fails to Override," *New York Times*, 28 June 1973, www.proquest.com.

3. 119 Cong. Rec. 22304-22325, 22336-22364 (29 June 1973).

Stabbed in the Back

1. Nixon, *No More Vietnams*, 165.

2. Nixon, *RN*, 888. Haig offers an alternative explanation in *Inner Circles*, which blames it all on one big misunderstanding. In Haig's version, he, not Nixon, talks by phone to Ford before the big vote:

> Ford, as honest as the day is long, made no effort to hide his surprise and dismay. He said he had been led to believe by Mel Laird that this bill was acceptable to the President. I told him that Laird's support was news to the President. Ford was stunned. If he reversed himself now, he might have to resign as Republican leader.
>
> Despite my concern, Nixon could not bring himself to ask Ford to do that: "Al, I can't afford to lose Jerry Ford."

According to Haig's account, Nixon would rather lose Vietnam than the House minority leader (see Haig, *Inner Circles*, 316–17). As we have seen, White House records show that the decisive phone conversation was between Ford and Nixon, not Ford and Haig, and as we shall see, Ford and Laird thought Nixon had the votes in the House to sustain a veto.

3. "Laird/Kissinger," 28 June 1973, Kissinger Telcons Box 20, RMNL, http://gateway.proquest.com/openurl?url_ver=Z39.88-2004&res_dat=xri:dnsa&rft_dat=xri:dnsa:article:CKA10348.

4. "House Again Clears Bill to Bar Bombing, This One 'Veto-Proof,' " *Wall Street Journal*, 27 June 1973, www.proquest.com.

5. John Herbers, "Test Intensifies; Rider Put on Another Bill after the House Fails to Override," *New York Times*, 28 June 1973, www.proquest.com.

6. Richard L. Madden, "Nixon Agrees to Stop Bombing by U.S. in Cambodia by Aug. 15, With New Raids up to Congress," *New York Times*, 30 June 1973, www.proquest.com.

7. Ellsberg, *Secrets*, 453.

"We Can Blame Them for the Whole Thing"

1. Conversation 424-026, 29 March 1973, 10:50 a.m., Executive Office Building.

2. "Melvin Laird/Kissinger," 26 June 1973, Kissinger Telcons Box 20, RMNL, http://gateway.proquest.com/openurl?url_ver=Z39.88-2004&res_dat=xri:dnsa&rft_dat=xri:dnsa:article:CKA10340.

3. "Letter to the Speaker of the House and the Majority Leader of the Senate about the End of United States Bombing in Cambodia," 3 August 1973, *PPPUS: Nixon, 1973*, www.presidency.ucsb.edu/ws/?pid=3929.

4. John W. Finney, "Nixon Sees Peril in Bombing Halt; Warns Congress," *New York Times*, 4 August 1973, www.proquest.com.

5. Kissinger, *Years of Upheaval*, 359.

6. According to Kissinger's book, "Laird insisted that we had no choice if we wanted the government to continue to function." According to Kissinger's telcon, Laird said the opposite regarding a combat ban for countries of Indochina other than Cambodia (Kissinger, *Years of Upheaval*, 359; "Laird/Kissinger," 28 June 1973, Kissinger Telcons Box 20, RMNL, http://gateway.proquest.com/openurl?url_ver=Z39.88-2004&res_dat=xri:dnsa&rft_dat=xri:dnsa:article:CKA10348).

Nixon's *Dolchstoßlegende*

1. David E. Sanger, "Panel Urges Basic Shift in U.S. Policy in Iraq," *New York Times*, 7 December 2006, www.proquest.com. Members included James A. Baker III, Lee H. Hamilton, Sandra Day O'Connor, Edwin Meese III, Vernon E. Jordan, William J. Perry, and Alan K. Simpson.

2. Michael R. Gordon, "Will It Work on the Battlefield?," *New York Times*, 7 December 2006, www.proquest.com.

3. Nixon, *No More Vietnams*, 165.

4. Berman, *No Peace, No Honor*, 8–9.

Unearthing Nixon's Strategy

1. Jeffrey Kimball, "Fighting and Talking," *Diplomatic History* 27, no. 5 (2003): 763–66.

2. Pierre Asselin, "Featured Review: Kimball's Vietnam War," *Diplomatic History* 30, no. 1 (2006): 163–67.

3. Ibid., 164 (both quotations).

4. Ibid., 165.

5. "Memorandum of Conversation," 9 July 1971, 4:35–11:20 p.m., attached to Lord to Kissinger, "Memcon of Your Conversations with Chou En-lai," 29 July 1971, NSCF Box 1033, RMNL.

6. Kissinger used "reasonable interval" and "sufficient interval" in "Memorandum of Conversation," 21 June 1972, 3:25–6:45 p.m., "China—Dr. Kissinger's Visit, June 1972 Memcons" folder, NSCF Kissinger Office Files Box 97, RMNL.

7. For the evidence that Kissinger negotiated a "decent interval" settlement, see Kimball, *The Vietnam War Files*, 186–91, 322n19.

8. Asselin, "Featured Review," 165.

9. Ibid., 165n3.

10. Kimball, *The Vietnam War Files*, 192.

11. Pierre Asselin, *A Bitter Peace: Washington, Hanoi, and the Making of the Paris Agreement* (Chapel Hill: University of North Carolina Press, 2002), 169.

12. John Carland, "A Roundtable on Richard Nixon and the Vietnam War," *Passport: The Society for Historians of American Foreign Relations Review* 43, no. 3 (January 2013): 23.

13. I find the *FRUS* volumes, by and large, enormously helpful overall, as anyone who checks my notes can see.

14. H-DIPLO, the H-NET network on Diplomatic History and International Affairs, published Kimball's review, "H-Diplo FRUS Review No. 5—*Foreign Relations of the United States, 1969–1976: Vietnam, January–October 1972. Volume VIII*," on 23 November 2011.

15. *FRUS 1969–1976*, 8: Document 282.

16. Conversation 793-006, 6 October 1972, 9:30–10:03 a.m., Oval Office.

17. *FRUS 1969–1976*, 8: Document 284.

18. Ibid., 8: Document 5.

19. Carland, "A Roundtable on Richard Nixon and the Vietnam War," 23.

20. Ibid.

21. "Memorandum of Conversation," 9 July 1971, 4:35–11:20 p.m., attached to Lord to Kissinger, "Memcon of Your Conversations with Chou En-lai," 29 July 1971, NSCF Box 1033, RMNL; *FRUS 1969–1976*, 14: Document 271.

22. "Memorandum of Conversation," 20 June 1972, 2:05–6:05 p.m., "China—Dr. Kissinger's Visit, June 1972 Memcons" folder, NSCF Kissinger Office Files Box 97, RMNL.

The University of Virginia's Miller Center

1. Conversation 451-023, 18 February 1971, 6:16–6:37 p.m., Oval Office.

2. Conversation 471-002, 19 March 1971, 7:03–7:27 p.m., Oval Office.

3. "Address to the Nation on the War in Vietnam," 3 November 1969, *PPPUS: Nixon, 1969*, www.presidency.ucsb.edu/ws/?pid=2303.

4. "Panel Interview at the Annual Convention of the American Society of Newspaper Editors," 16 April 1971, *PPPUS: Nixon, 1971*, www.presidency.ucsb.edu/ws/?pid=2982.

5. "Second Annual Report to the Congress on United States Foreign Policy," 25 February 1971, *PPPUS: Nixon, 1971*, www.presidency.ucsb.edu/ws/?pid=3324.

6. "The President's News Conference," 30 July 1970, *PPPUS: Nixon, 1970*, www.presidency.ucsb.edu/ws/?pid=2603.

Decision Points

1. Conversation 475-016, 8 April 1971, 9:18–10:07 a.m., Oval Office.

2. There's some ambiguity over exactly when Nixon and Kissinger first proposed a ceasefire-in-place. Kimball notes correctly that in the September 1970 secret negotiations with the North, Kissinger "had not directly called for North Vietnamese troop withdrawal, which implied that Washington was abandoning its longstanding demand for mutual withdrawal" (Kimball, *The Vietnam War Files*, 133–34). After that session, however, Nixon proposed a different kind of ceasefire-in-place in October 1970—an immediate one in which American and allied troops—as well as the North and South Vietnamese—would remain in place while an Indochina Peace Conference worked out a settlement. "I propose that all armed forces throughout Indochina cease firing their weapons and remain in the positions they now hold. This would be a 'cease-fire-in-place.' It would not in itself be an end to the conflict, but it would accomplish one goal all of us have been working toward: an end to the killing," Nixon announced in a televised address ("Address to the Nation about a New Initiative for Peace in Southeast Asia," 7 October 1970, *PPPUS: Nixon, 1970*, www.presidency.ucsb.edu/ws/?pid=2708). Nixon said nothing about abandoning his demand for North Vietnamese withdrawal in return for American withdrawal. In fact, the next day he reaffirmed his demand for mutual withdrawal in the final settlement, telling reporters: "We made this proposal because we wanted to cover every base that we could. And so, that is why we offered the cease-fire, a total cease-fire. That is why we offered a total withdrawal of all of our forces, something we have never offered before, *if we had mutual withdrawal on the other side*" ("Replies to Reporters' Questions about Reaction to Address on Southeast Asia," 8 October 1970, *PPPUS: Nixon, 1970*, italics added, www.presidency.ucsb.edu/ws/?pid=2710). Nixon clarified once again that he was calling for mutual withdrawal in his next televised address: "I am sure most of you will recall that on October 7 of last year in a national TV broadcast, I proposed an immediate cease-fire throughout Indochina, the immediate release of all prisoners of war in the Indochina area, an all-Indochina peace conference, the *complete withdrawal of all outside forces*, and a political settlement" ("Address to the Nation on the Situation in Southeast Asia," 7 April 1971, *PPPUS: Nixon, 1971*, italics added, www.presidency.ucsb.edu/ws/?pid=2972). Kissinger wrote that in the 31 May 1971 proposal, "We gave up our demand for mutual withdrawal, provided Hanoi agreed to end all additional infiltration into the countries of Indochina," and "proposed . . . a ceasefire-in-place throughout Indochina to become effective at the time when US withdrawals began" (Kissinger, *White House Years*, 1018). More specifically, they formally proposed a ceasefire-in-place ("Memorandum of Conversation," 31 May 1971, 10:00 a.m.–1:30 p.m., "Camp David—Vol. VII" folder, NSCF Box 853, RMNL).

3. Conversation 507-004, 29 May 1971, 8:13–10:32 a.m., Oval Office.

4. Conversation 532-011, 30 June 1971, 10:18–10:30 a.m., Oval Office.

5. Nixon, *RN*, 689. "Hanoi," Kissinger wrote, "was obviously in full retreat from its hitherto unyielding position that Thieu would have to resign *before* anything else happened" (Kissinger, *Ending the Vietnam War*, 307).

6. Conversation 760-006, 3 August 1972, 8:28–8:57 a.m., Oval Office.
7. Ibid.
8. Conversation 789-006, 30 September 1972, 10:56 a.m.–12:00 p.m., Oval Office.
9. Nixon wrote that Hanoi's 8 October 1972 proposals "amounted to a complete capitulation by the enemy: they were accepting a settlement on our terms" (Nixon, *RN*, 692). "Le Duc Tho's draft agreement," Kissinger wrote, "in effect accepted what we had put forward on May 31, 1971, and on May 8, 1972" (Kissinger, *White House Years*, 1347).
10. Kissinger, *Ending the Vietnam War*, 329.

The Nixon Tapes

1. Douglas Brinkley and Luke Nichter, eds., *The Nixon Tapes: 1971–1972* (Boston: Houghton Mifflin Harcourt, 2014), 627.
2. Conversation 793-006, 6 October 1972, 9:30–10:03 a.m., Oval Office.
3. Ibid.
4. Ibid.
5. Brinkley and Nichter, *The Nixon Tapes*, 9–10.
6. Jeffrey Kimball, "Decent Interval or Not? The Paris Agreement and the End of the Vietnam War," *Passport: The Society for Historians of American Foreign Relations Review* 34, no. 3 (December 2003): 26–31.
7. Brinkley and Nichter, *The Nixon Tapes*, 152–57.
8. Ibid., 16.
9. Conversation 451-023, 18 February 1971, 6:16–6:37 p.m., Oval Office.
10. Del Quentin Wilber, "Nixon Fixation Pushes Professor to Listen to All Tapes," *Bloomberg Businessweek*, 9 August 2014, www.bloomberg.com/news/2014-08-08/nixon-fixation-pushes-professor-to-listen-to-all-tapes.html.

Interpretive Inertia

1. Passing mentions of the phrase in newspapers indicate that it was not unknown, but they do not define its meaning. "Assuming [a settlement] comes unstuck, does it matter how soon—what would be a decent interval between our departure and disaster?" asked one editorial ("Vietnam: The Missing Ingredient," *Washington Post*, 5 April 1970, www.proquest.com). "As the geopoliticians sometimes put it," Robert G. Kaiser wrote, "could the Americans withdraw and leave behind a decent interval before fate took its course in South Vietnam?" (*Washington Post*, 31 May 1970, www.proquest.com).
2. Frank Snepp, *Decent Interval: An Insider's Account of Saigon's Indecent End Told by the CIA's Chief Strategy Analyst in Vietnam* (New York: Random House, 1977), 50.
3. Kissinger, *White House Years*, 1359.
4. Nixon, *No More Vietnams*, 103.
5. Bundy, *A Tangled Web*.
6. Jeremi Suri, *Henry Kissinger and the American Century* (Cambridge: Belknap Press of Harvard University Press, 2009), 227.
7. Ibid., 232–33.

8. Conversation 760-006, 3 August 1972, 8:28–8:57 a.m., Oval Office.

9. "Memorandum of Conversation," 9 July 1971, 4:35–11:20 p.m., attached to Lord to Kissinger, "Memcon of Your Conversations with Chou En-lai," 29 July 1971, NSCF Box 1033, RMNL; "Memorandum of Conversation," 22 February 1972, 2:10–6:10 p.m., President's Office Files Box 87, RMNL; "Memorandum of Conversation," 20 June 1972, 2:05–6:05 p.m., "China—Dr. Kissinger's Visit, June 1972 Memcons" folder, NSCF Kissinger Office Files Box 97, RMNL; "Memorandum of Conversation," 21 June 1972, 3:25–6:45 p.m., "China—Dr. Kissinger's Visit, June 1972 Memcons" folder, NSCF Kissinger Office Files Box 97, RMNL; "Memorandum of Conversation," 22–23 June 1972, 11:03 p.m.–12:55 a.m., "China—Dr. Kissinger's Visit, June 1972 Memcons" folder, NSCF Kissinger Office Files Box 97, RMNL.

10. Conversation 793-006, 6 October 1972, 9:30–10:03 a.m., Oval Office; *FRUS 1969–1976*, 14: Document 271.

11. Nixon, *No More Vietnams*, 165. Kissinger likewise bemoaned the "consequences . . . for America's reputation as a reliable ally" of the Nixon-invited congressional ban on American combat in Indochina (Kissinger, *Years of Upheaval*, 360).

12. Niall Ferguson, "The Jewish Key to Henry Kissinger," *Times Literary Supplement*, 28 May 2008, www.the-tls.co.uk; Joe Hagan, "The Once and Future Kissinger," *New York*, 26 November 2006, http://nymag.com; Jacob Heilbrunn, "Got Your Back," *New York Times*, 19 July 2009, www.nytimes.com.

13. In *Nixon and Kissinger: Partners in Power*, Robert Dallek is vague on the subject, writing, "To ensure against subsequent complaints of failure in Vietnam, which seemed certain to follow a quick Saigon collapse after a U.S. departure, Nixon and Kissinger hoped Hanoi would allow a 'decent interval' before it toppled Thieu's government" (Robert Dallek, *Nixon and Kissinger: Partners in Power* [New York: HarperCollins, 2007], 620). They did more than hope: they provided secret assurances (of nonintervention if Hanoi gave them an interval of about eighteen months) and made implicit threats (of intervention if Hanoi tried to overthrow Saigon too quickly). Dallek includes bits and pieces of evidence, but for some reason he leaves out the most compelling and illuminating passages from Kissinger's negotiations with Zhou in July 1971 and June 1972.

Last Days in Vietnam

1. *Last Days in Vietnam*, documentary, directed by Rory Kennedy, Moxie Firecracker Films.
2. Ibid.
3. Ibid.
4. "Address Before a Joint Session of the Congress Reporting on United States Foreign Policy," 10 April 1975, *Public Papers of the Presidents of the United States: Gerald R. Ford, 1975* (Washington, DC: GPO, 1977), www.presidency.ucsb.edu/ws/?pid=4826 (hereafter *PPPUS: Ford, 1975*).
5. John H. Averill, "Congress Votes Viet Aid, Use of Troops," *Los Angeles Times*, 24 April 1975, www.proquest.com.

6. United Press International, "Conferees Agree on $327 Million Evacuation Bill," *Los Angeles Times*, 25 April 1975, www.proquest.com.

7. Douglas E. Kneeland, "Goal Put at 8,000 a Day," *New York Times*, 25 April 1975, www.proquest.com.

8. *Last Days in Vietnam*, documentary.

9. "Remarks and a Question-and-Answer Session at the Annual Convention of the American Society of Newspaper Editors," 16 April 1975, *PPPUS: Ford, 1975*, www.presidency.ucsb.edu/ws/?pid=4837.

10. Murrey Marder, "Last Exit," *Washington Post*, 11 April 1975, www.proquest.com.

11. Louis Harris, "Harris Finds Most Against Troop Use," *Washington Post*, 24 April 1975, www.proquest.com.

12. Ann Hornaday, "'Last Days in Vietnam' Movie Review: The Stories behind the Iconic Images," *Washington Post*, 11 September 2014, http://wapo.st/1qHC3f9.

13. George Packer, "Obama and the Fall of Saigon," *New Yorker*, 10 September 2014, www.newyorker.com/news/daily-comment/obama-fall-saigon.

14. Douglas E. Kneeland, "Goal Put at 8,000 a Day," *New York Times*, 25 April 1975, www.proquest.com.

15. Cong. Rec. 18,783 (8 June 1971) (quoted in statement of Sen. Philip Hart).

16. R. W. Apple Jr., "Nixon Would Bar Forced Coalition in South Vietnam," *New York Times*, 28 October 1968, www.proquest.com.

A Better War

1. Sorley, *A Better War*, 217.
2. Ibid., 218–19.
3. Ibid., 161.
4. Lewis Sorley, *Thunderbolt: General Creighton Abrams and the Army of His Times* (Bloomington: Indiana University Press, 1992), 343, italics added.
5. Sorley, *A Better War*, 357, italics added.
6. "Remarks in Detroit at the Annual Convention of the Veterans of Foreign Wars," 19 August 1968, *Public Papers of the Presidents of the United States: Lyndon B. Johnson, 1968* (Washington, DC: GPO, 1970), www.presidency.ucsb.edu/ws/?pid=29085.
7. Conversation 823-001, 14 December 1972, 9:59–11:46 a.m., Oval Office.
8. Peter Spiegel and Jonathan Weisman, "Behind Afghan Debate, a Battle of Two Books Rages," *Wall Street Journal*, 7 October 2009; Fred Kaplan, "We Still Don't Have a Plan," *Slate*, 16 November 2005, slate.com; David Ignatius, "A Better Strategy for Iraq," *Washington Post*, 4 November 2005, www.washingtonpost.com.

The Aid-Cutoff Myth

1. Melvin R. Laird, "Iraq: Learning the Lessons of Vietnam," *Foreign Affairs* 84, no. 6 (November–December 2005): 22–43, www.jstor.org/stable/info/20031774.
2. Nixon, *No More Vietnams*, 193; Sorley, *A Better War*, 367.

3. Laird used the term "Iraqization" in an interview with James Glanz, "Saving Face and How to Say Farewell," *New York Times*, 27 November 2005, www.nytimes.com.

4. Liz Sly and Ahmed Ramadan, "Insurgents Seize Iraqi City of Mosul as Security Forces Flee," *Washington Post*, 10 June 2014; Loveday Morris and Liz Sly, "Iraq Disintegrating as Insurgents Advance toward Capital; Kurds Seize Kirkuk," *Washington Post*, 13 June 2014, www.washingtonpost.com.

5. Martin Chulov, Fazel Hawramy, and Specer Ackerman, "Iraq Army Capitulates to Isis Militants in Four Cities," *Guardian* (United Kingdom), 11 June 2014, www.theguardian.com.

6. Eric Schmitt and Michael R. Gordon, "The Iraqi Army Was Crumbling Long before Its Collapse, U.S. Officials Say," *New York Times*, 12 June 2014, www.nytimes.com.

How Wars Don't End

1. Gideon Rose, "What Would Nixon Do?" *New York Times*, 25 June 2011, www.nytimes.com/2011/06/26/opinion/sunday/26afghan.html. The *Times* replaced "The Week in Review" with the "Sunday Review" on 26 June 2011 (The Editors, "To Our Readers," *New York Times*, 25 June 2011, www.nytimes.com/2011/06/26/opinion/sunday/26note.html). I once started drafting a humor piece called "What Would Nixon Do?" I meant it ironically, in the vein of *The Screwtape Letters* by C. S. Lewis, but Rose was seriously advising Obama to follow Nixon's example.

2. The Miller Center released the transcript in connection with the thirtieth anniversary of Nixon's resignation (Associated Press, "Tape: Nixon Mulled Vietnam Exit in 1972," *USA Today*, 8 August 2004, http://usatoday30.usatoday.com/news/washington/2004-08-08-nixon_x.htm).

3. Ken Hughes, "Fatal Politics: Nixon's Political Timetable for Withdrawing from Vietnam," *Diplomatic History* 34, no. 3 (June 2010): 497–506.

Questions Unasked

1. Associated Press, "US Review Finds Iraq Deadlier Now Than a Year Ago, as Officials Weigh Extending Troop Presence," *Washington Post*, 30 July 2011, www.washingtonpost.com; Special Inspector General for Iraq Reconstruction, "Quarterly Report and Semiannual Report to the United States Congress," 30 July 2011, http://cybercemetery.unt.edu/archive/sigir/20130930185834/http://www.sigir.mil/publications/quarterlyreports/index.html.

2. Carl Hulse, "Pelosi Cautions Bush Not to Veto an Iraq Bill," *New York Times*, 11 March 2007, www.nytimes.com.

3. Glenn Kessler, "White House Postponing Loss of Iraq, Biden Says," *Washington Post*, 5 January 2007, www.washingtonpost.com.

4. "An American Century: A Strategy to Secure America's Enduring Interests and Ideas; A Romney for President White Paper," 7 October 2011, Romney for President, Inc., www.MittRomney.com.

5. Ibid.

6. James Dao and Andrew W. Lehren, "In Toll of 2,000, New Portrait of Afghan War," *New York Times,* 21 August 2012, www.nytimes.com. In 2014, Obama announced plans to keep approximately 10,000 American soldiers in Afghanistan past the end of that year and to reduce the American presence by the end of 2016 to a small force protecting the embassy and handling other security matters (Mark Landler, "U.S. Troops to Leave Afghanistan by the End of 2016," *New York Times,* 27 May 2014, www.nytimes.com).

INDEX

ABC Evening News: Kissinger's "Peace Is at Hand" briefing covered, 131; Nixon's speech carried by, 2; soldier's son's story on, 21; on Thieu's demands, 126–27, 138; on Vietnam settlement nearly complete, 158

Abrams, Creighton W.: bombing offensives and, 58–59, 61, 63, 68, 69; failure of, 198–99; negotiations with Saigon, 113–14, 123; Nixon on, 67, 68; Nixon's order countermanded by, 58–59; on South Vietnam's inability to survive, 14–15, 49, 72, 198, 205, 228n31; strategies of, 59–61, 200

Afghanistan War (2001–): "Nixonian approach" in, 202–6; strategies used in, 200; unasked questions about, 206–9; withdrawal of troops from, 254n6

Agnew, Spiro T., 60, 67, 68, 139

aid-cutoff myth, 200–202

Ambrose, Stephen E., 215n9

American POWs. *See* POWs

American Society of Newspaper Editors, 24–25

American troops: as advisers, x, 25–26, 59–60, 61–62, 162, 204–5; call for bombing vs. boots on the ground, 64–66; commander in Vietnam (*see* Abrams, Creighton W.); costs of Nixonian approach for, xii, 86, 196–97, 200–209; dilemmas concerning, after Vietnam settlement signing, 161–63, 245–46n5; educational benefit increased for, 126; effects of prolonged war on, 7–8; Ted Kennedy on Nixon's disregard for, xi, 45, 49, 82–83; Nixon's actual position on, 27–28; Nixon's public position on, 17–18, 181, 189; "search and destroy" vs. "clear and hold" strategies of, 200, 207; in unwinnable war, 189, 191–92, 199. *See also* casualties, US; POWs; Vietnamization; withdrawal of troops

Anderson, Jack, 229n44

anti-Communism: liberals' belief in Nixon's extreme, 49–54; Madman Theory linked to, 163, 174, 245–46n5; "Peace Is at Hand" briefing as vindication of, 134

antidrug campaign, 4

anti-Semitism, 2–3

antiwar movement: attitudes toward losing war in, 23–24; Beatles in, 4; blamed for defeat, 74–75; Hanoi's link to, 7; Nixon on, 143; Nixon's credibility and, 71

"any means necessary" message, 70–74, 230n3

Arlington National Cemetery, Veterans Day ceremony, 124

Army of the Republic of [South] Vietnam (ARVN): advances of, 69; breaking under Easter Offensive, 57–58; inability to survive without US support, 14–15, 27, 49, 72, 198, 205, 228n31, 229n44; incursion into Laos, 11, 12–14; Nixon's refusal to discuss collapse of, 76; number of troops, 57, 114, 117; one-to-one demobilization of North and (proposed), 147; South Vietnamese

Army of the Republic of [South] Vietnam (ARVN) (*continued*) leaders' view of, 233n6; US advisers and training of, x, 25–26, 59–60, 61–62, 162, 204–5; US resupply of armaments for, 99, 105, 111, 112, 152. *See also* Saigon government; Vietnamization

Arrogance of Power, The (Fulbright), 155

As I Saw It (Rusk), 46

backstabbing myth: actual events vs. general belief in, 170–73; bombing restrictions alleged in, 65, 66; cost of continued belief in, 86, 196–97, 200–209; documentary's embellishment of, 192–96; evidence that overturns, xii, 173–74, 176, 178, 186, 189; foundation for, 74–75; Nixon's fabrication of, ix, xii, 167–70, 172, 188, 191–92; set up for, 163–67; suppression of evidence key to, 173

Bay of Pigs Invasion (1961), 37, 39, 40

Berman, Larry, 172–73

Bernstein, Carl, 108, 128

Best and the Brightest, The (Halberstam), 50

Better War, A (Sorley), 59, 197–200

Biden, Joe, 207

"Big" Minh (Duong Van Minh), 140, 233n7

Bitter Peace, A (Asselin), 174–76

bombing halt (under Johnson), 70, 98, 132, 140–41, 224n8

bombing offensives (1972–1973): all-Indochina combat ban of future, 163–70; in Cambodia, 163–64, 167–68; Christmas surge in (Linebacker II), 150–54, 159, 160, 244n3, 245n5; context of, 55–57, 64–66, 70–74; continued during Vietnam settlement, 124, 129, 137, 150–52; Democrats' misreading of, 127, 172–73; dispute concerning, 58–59, 67–69; impossibility of, after signing Vietnam settlement, 161–63; later reflections on, 70; May 8 surge (Linebacker I), 56–58, 62–63, 69, 76–77; Nixon's claims for success, 69, 76, 230n3; Nixon's micromanagement of, 154; Nixon's rhetoric on, 20; overestimation of results, 96; as political cover for "decent interval" negotiations, 77–78, 150; polling on, 55–56, 58, 64, 69, 71–73, 118, 124, 150, 159, 244n3, 245n5; possibility of continuing post-election and into 1973, 89, 91, 120–21; post-election resumption of, 150–52; restrictions on, 65–66, 67–68; results of, 62–64, 69, 228n38; targets in, 59–61; Thieu's rejection of deal as reason for, 155. *See also* mining North's harbors

Brando, Marlon, 149

Brezhnev, Leonid, 76, 77

Brinkley, Douglas, 185–87

Brookings Institution, 50. *See also* Chennault, Anna

Brown, L. Dean, 195

Buckley, James, 133

Buckley, William F., 133

Bundy, McGeorge, 134

Bundy, William, 189

Bunker, Archie (television character), 81

Bunker, Ellsworth, 89, 93, 107

Bureau of Narcotics and Dangerous Drugs, 4

Burr, William, 230n6

Bush, George H. W., 31, 133

Bush, George W., 201, 206–8, 209. *See also* Afghanistan War; Iraq War

California Central Republican Committee, 37

Cambodia: blame for fall to Communists, 167–70; bombing in, 163–64, 167–68; China's concern about US in, 32; effects of secret bombing in, 11–12, 163–64; Nixon's concern about collapse of, 29–30; "security

guarantee" concerning Ho Chi Minh Trail and, 94, 98–99, 100, 113, 116, 117, 153; US ban on bombing of, 164–66; US incursion into, 12, 50; Vietnam settlement backed by, 123. *See also* Ho Chi Minh Trail
Capen, Richard, 6
Carland, John M., 176–79, 248n13
Castro, Fidel, x, 39–40, 42, 168
casualties, US: in Cambodian incursion, 12; decried by McGovern, 102, 127; in hiding Nixon's failed strategy, 170–73; Nixon's calculations of, 72; in North's Easter Offensive, 57, 76, 102; number during Nixon's presidency, 87, 92, 108, 127, 133, 142; in rejecting coalition government, 44
casualties, Vietnamese, 14, 160. *See also* fall of Saigon
CBS Evening News: Kissinger's "Peace Is at Hand" briefing covered, 131; McGovern on, 101; Nixon's speech carried by, 2; soldier's son's story on, 21; on Thieu's demands, 126–27, 138; on Vietnam settlement nearly complete, 158
CBS *Face the Nation*, 65
CBS Morning News, 127
ceasefire-in-place: bombing options and, 28; concept proposed, 118, 135, 183; face-saving withdrawal of some NVA troops from South (proposed), 94–95, 120–21, 125, 147; infiltration likely to continue in, 29, 73, 98, 99, 100, 122, 153; length of "decent interval" before breaking, 35, 77, 78, 179; mutual withdrawal not excluded by, 249n2; negotiations concerning, 26–30; NVA troops to remain in South during, 106, 113, 114, 116–17, 118, 183, 193, 198–99, 200, 232n4; South's "chance" of survival in, 15–16, 120–22, 180–81, 197–98, 206; Thieu's objections to North's troops in South during, 117, 125, 127,

136–40, 183; violations of, after signing Vietnam settlement, 160
Central Intelligence Agency. *See* CIA
Chancellor, John, 26, 36
Chapin, Dwight, 108
Chennault, Anna (and Chennault Affair), 140–41, 149, 224n8
Chicago Tribune, 36, 139
China: attitude toward US, 97; blame for loss to Communists, x, 168; "decent interval" strategy and role of, 30–36, 54, 78–81, 85, 118, 162, 174–76, 190; hostility with Soviet Union, 31, 34; intervention in war, 32, 62–63, 76–77, 220n8; Kissinger's secret talks with, 30–36, 54, 78–81, 174–76, 190; Nixon's announcement of visit to, 36–37; Nixon's visit to, 54–55, 64; US ping-pong team invited to, 21; US State Dept. study on, 38–39; Vietnam negotiations and, 90, 123, 124–26. *See also* Mao Zedong; Zhou Enlai
China White Paper, 38–39
Churchill, Winston, 91
CIA (Central Intelligence Agency): on bombing, 63; on South Vietnam's inability to survive, 48–49, 72, 198, 205; on spring offensive results, 62
Clifford, Clark, 50, 52
coalition government (later dropped): as "disguised surrender," 44, 98, 119, 127, 197; Hanoi's demand for, 44, 80, 81, 83, 99, 130; Kissinger's 1969 *Foreign Affairs* article and, 135; Nixon's rejection of, 44, 119, 128, 132, 238n1; Thieu's resignation tied to, 183
Cold War politics: "decent interval" strategy in context of, x–xi; rhetoric of blame in, x, 29, 37–42, 48, 168; US geopolitical credibility forfeited in, 97, 190–91, 203. *See also* China; Soviet Union; United States
Collier, Bernard Law, 37
colonial legacy, 102. *See also* France

Colson, Charles W. "Chuck," 102, 103, 131, 132, 133
Committee for the Re-Election of the President (CREEP), 101, 128
Communism: blame for losing countries to, x, 29, 37–42, 48, 168. *See also* China; Cuba; North Vietnam; Soviet Union
Communist takeover of South Vietnam: Nixon's actual view on timing of, 3–4, 7, 28–30, 108, 124, 176, 185, 203–5; Nixon's publicly declared rejection of, 25–26, 128, 129; public opposition to, 23–24. *See also* "decent interval" strategy
confirmation bias, 173
Cooper, Gary, 148
Coppola, Francis Ford, 149
Council on Foreign Relations. See *Foreign Affairs*
CREEP (Committee for the Re-Election of the President), 101, 128
Cronin, Paul, 165
Cuba: blame for loss to Communists, x, 39–42, 168; interventions in Africa, 190
Cuban Missile Crisis (1962), 37, 38

Dailey, Peter, 101
Dallek, Robert, 251n13
Dean, John, 166–67
Decent Interval (Snepp), 187–88
"decent interval" strategy: absolute secrecy of, 95, 102, 153; assurances of US nonintervention in North's takeover of South in, xii, 30, 35, 77–78, 80, 86, 97, 114–15, 117, 138–39, 146, 153, 162, 169, 175–76, 179, 180, 184, 190–91, 203, 251n13; bombing's role in, 77–78, 150; China's agreement to, 81; context and details of, x–xi, 30; cost of, 86, 196–97, 200–209; evidence on, 173–76, 178–79, 180–81, 186–92; Humphrey's speech in relation to, 47–48; immorality of, x, 215n4; JFK's use of, 45–46;

Kissinger's negotiations with China tied to, 30–36, 78–81; Kissinger's scribbled note on, 30, 174; media mentions of, 250n1; Nixon and Kissinger's strategizing on timing for, 84–86, 160–61, 184–85; Nixon's motivation for, 3–4, 7, 28–30, 108, 124, 176, 180, 185, 203–5; North's upholding of, 193; opposition to, 53–54; POWs' families and, 9–10, 107–9; prisoners dilemma linked to, 161–63; question of Bush's use of, 207–8; settlement needed to ensure, 91; South's eventual destruction inherent in, 77–78, 89–90, 92, 93, 112; specific timing for, 35, 77, 78, 179; as surrender, 44, 150, 162, 196–97; synonyms for, 79–80, 87, 99–100; Thieu needed in place for successful, 93, 100, 105–6, 115–16, 122; Thieu's blocking of Vietnam settlement and, 119–25
Dedmon, Emmett, 25
Defense Intelligence Agency (DIA), 62
demilitarized zone (DMZ), 56, 71, 141, 147, 152–53, 193
Democrats: blamed for bombing failure in 1960s, 64–66; blamed for losing countries to Communism, x, 29, 37–39, 168; Bush's actual strategy possibly misread by, 207; congressional election results (1972), 145; Nixon and Kissinger's plan to destroy party, 19–20, 30, 84, 104, 109, 121, 123, 131, 167; Nixon's actual strategy misread by, xi–xii, 49–54, 101–3, 141–42, 151, 169, 172–73; Nixon's opening of China and, 33; "Peace Is at Hand" briefing response of, 134–35; retrenching in positions of, 156–57; self-sabotage of, 81–83, 139; warned about setting date of withdrawal, 16–17; what-ifs of troop pullout and, 47–49; withdrawal-for-POWs deal proposed by, 154–56. *See also* liberals

Denton, Jeremiah, 162–63
DIA (Defense Intelligence Agency), 62
Dickinson, William, 24
Diem (Ngo Dinh Diem), 92, 138, 233n7
Diplomatic History (journal), 174, 203, 204
DMZ (demilitarized zone), 56, 71, 141, 147, 152–53, 193
Dobrynin, Anatoly, 33, 70, 74, 75, 119–20, 122–23, 124
documentary on fall of Saigon, 192–97
Dolchstoßlegende, post-WWI Germany, ix. *See also* backstabbing myth
draft dodgers, 110
Dubik, James, 201
Duc (Nguyen Phu Duc), 116, 148–49, 155, 156
Duong Van "Big" Minh, 140, 233n7

Eagleton, Thomas, 82, 83
Eastern Europe, blame for loss to Communists, x, 37, 168
Ehrlichman, John, 50, 129
Eisenhower, Dwight D.: blamed for foreign policy weakness, 39–41; Korean War decisions of, 27, 37, 70; secret tapes of, 179; Vietnam War decisions of, 53, 132
election commission for South Vietnam. *See* National Council of National Reconciliation and Concord
elections, US
—1946: Republican successes, 168
—1960 (Kennedy/Nixon): campaign rhetoric, 39–43; narrow victory, 98
—1968 (Humphrey/Nixon): LBJ, Chennault Affair, and, 140–41, 149, 224n8; Thieu's blowup in, 97–98, 132, 140–41
—1972 (McGovern/Nixon): Democratic self-sabotage in, 81–83, 90; keeping negotiation details confusing before, 120–21, 123; McGovern's campaigning, 82–83, 90, 127, 132, 139, 141–42; McGovern's concession speech, 145; Nixon and Kissinger's August 1972 strategizing for, 83–86; Nixon's actual vs. depicted actions in, 146–47; Nixon's campaigning, 103–4, 107–10, 122, 124, 128, 142–43; Nixon's decision to time troop withdrawal to, x, 3–5, 7–8, 10–11, 15–16, 25–26, 28, 48–49, 179–81, 182–83, 186, 203; Nixon's denial of Vietnam settlement or withdrawal date linked to, 91–92, 111, 112, 115; Nixon's landslide victory, 143–45; Nixon's nomination acceptance, 86–87; Nixon's nonpartisan pose, 142–43, 241n2; Nixon's secret espionage fund and, 108, 128; other politicians' link of withdrawal date to, xi, 22–23, 45–47, 49, 82; "Peace Is at Hand" briefing and, 130–33; polling on, 87, 98, 105, 122; President of Perpetual War rhetoric in, 52; residual troops idea rejected in concern for, 27; "saving face or saving lives" question in, 101–3, 127, 133, 141–42; Shriver's campaigning in, 138–39; spring offensive timing and, 55–56, 57–58; Thieu's blowup in, 136–40; Vietnam settlement not needed to win, 106–7, 115; Vietnam settlement preferred after, 122–26; vote count, 144
—1974 (midterms): strategizing "decent interval" for fall of Saigon and, 160–61
Ellsberg, Daniel: on JFK and Vietnam, 46; on Nixon's devotion to Saigon, 53–54, 166–67; NSSM-1 leaked by, 63, 229n44
Ervin, Sam, 166
Europa, L' (periodical), 148
Evans, Rowland, 70
evidence: aid-cutoff myth dismissed by, 200–202; backstabbing myth shattered by, xii, 173–74, 176, 178, 186, 189; on decent interval strategy,

evidence (*continued*)
173–76, 178–79, 180–81, 186–92; decision points and contradictions in, 181–85; interpretive inertia and, 187–92; limited awareness of, 176–77; omissions and suppression of, 173, 177–79, 185–87, 188, 189, 192–97, 251n13. *See also* secret recordings

Fallaci, Oriani, 148
fall of Saigon: assurances of US nonintervention in, xii, 30, 35, 77–78, 80, 86, 97, 114–15, 117, 138–39, 146, 153, 162, 169, 175–76, 179, 180, 184, 190–91, 203, 251n13; Biden's reference to, 207; blame for, ix, xii, 74–75, 169, 170–73; cost of, 196–97; documentary on, 192–97; expected by Nixon and planned by Hanoi, 7–8, 11, 49, 77–78, 85–86, 88, 89–95, 96–97, 112, 113–16, 138–39, 147, 185–86, 196–97, 198; implications for Christmas bombing, 160; Nixon and Kissinger's strategizing on timing for, 160–61, 183–85; Nixon's concern about, before 1972 election, 3–4, 7, 28–30, 90, 108, 124, 176, 185, 203–5; other politicians' attention to possibility of, 138–39; predicted recriminations for, 16–17, 20, 49; speed of, 124. *See also* backstabbing myth; "decent interval" strategy; Saigon government
FBI (Federal Bureau of Investigation), 50–51
Ferguson, Niall, 192
force majeure (greater force), 170
Ford, Gerald "Jerry" R.: all-Indochina combat ban and, 163–66, 169, 171, 246n2; call for bombing offensive, 65; last-minute aid requests of, 193–96; Nixon's aid-cutoff threat and, 149, 156
Foreign Affairs (journal): aid-cutoff myth in, 200, 201; Kissinger on Vietnam in, 135; "Nixonian approach" urged in Afghanistan by editor of, 202–6; Nixon on China in, 31
Foreign Assistance Appropriations Act (1975), 194
Foreign Relations of the United States (*FRUS* vols.), 176–79, 248n13
France: Algerian independence from, 95; diplomatic mission in Hanoi bombed, 104; negotiations and withdrawal from Indochina, 9, 55, 70; Nixon's visit to, 95–96. *See also* Geneva Accords
Frankel, Max, 106
French Communist Party, 99
Frishman, Robert, 6
FRUS (*Foreign Relations of the United States*) volumes, 176–79, 248n13
Fulbright, J. William, 155–56

Gallup, George (pollster), 2, 24, 47, 64–65, 98, 159, 245n5
Gaulle, Charles de, 55, 95–96
Gelb, Leslie H., 50–52
Geneva Accords (1954): DMZ in, 152; Kissinger on, 35; negotiations for, 9, 55, 70; Paris Accords compared with, 132–33
Geneva Conventions, 6–7
Germany, post-WWI *Dolchstoßlegende* in, ix. *See also* backstabbing myth
Giaimo, Robert, 164
Godfather, The (film), 149
Goldwater, Barry M.: call for bombing offensive, 65; Nixon's aid-cutoff threat and, 149, 156–58, 160, 176, 244n32
Gordon, Michael R., 170–71, 172
Graham, Billy, 110
Gravel, Mike, 63
Griffin, Robert, 22–23
Gritz, "Bo," 8–9
Gromyko, Andrei, 77–78, 88, 177, 205

Haig, Alexander: on all-Indochina combat ban, 246n2; despondency of,

104, 235n3; on incursion into Laos, 13; on Adm. McCain, 67; mentioned, 11, 31; Nixon's frankness with, 120–21, 150; in Nixon's meeting with Nguyen Phu Duc, 148–49; Nixon's speech praised by, 20; Saigon trip of, 89, 233n11; spring offensive plans of, 57, 58; on Vietnam settlement, 105–6, 113, 116, 151–52

Haiphong, mining of. *See* mining North's harbors

Halberstam, David, 50

Haldeman, H. R. "Bob": on "ad libbed" speech ending, 22; Kissinger's name for, 3, 215n6; on Kissinger's "Peace Is at Hand" briefing, 131; McGovern and, 102–3; mentioned, 21, 50; Nixon's campaign stops and, 103, 108–9; Nixon's Madman Theory and, 245–46n5; polls arranged and reported by, 24, 118; record keeping by, 3, 216nn7, 9; on timing of fall of Saigon, 160; Vietnam settlement discussed with, 104–6; Watergate linked to, 128; on withdrawal of troops, 3–5, 23

Halperin, Morton H., 50–52, 53, 166

Hammond, William, 60

Hanhimäki, Jussi, 221n7, 231n1

Hanoi (city): bombing targets in and around, 152; French diplomatic mission bombed, 104; restrictions to avoid bombing Soviet embassy during Moscow Summit, 68; US air reconnaissance of, 137

Hanoi government: attitude toward Nixon, 73–74, 163, 245–46n5; China's support for, 54–55; demand for Thieu's resignation dropped by, 99, 105, 183–84, 249n5; draft agreement release and, 121, 124, 128–30; expected takeover of South by, 7–8, 11, 49, 77–78, 85–86, 88, 89–95, 96–97, 112, 113–16, 138–39, 147 (*see also* fall of Saigon); infiltration strategy of, 29, 73, 98, 99, 100, 122, 153; Kissinger's secret talks with, 26–30, 54, 183; LBJ's ultimatum to, 71; Nixon's promises and threats to, 70–74, 147–48; October 31 deadline for signing Vietnam settlement, 98, 104, 125, 128, 136–37; Paris Accords as legitimating presence in South, 122; Politburo and, 81, 231n11; POWs tortured and mistreated by, 6–7; pressured to settle, 90; proposed withdrawal deadline extended by, 10; "security guarantees" requested from, 94, 98–99, 100, 113, 116, 117, 130, 153; settlement proposal of (Oct. 8, 1972), 99–100, 184–85, 250n9; supplies and flexibility of, 62–63; Thieu's blocking of Vietnam settlement and, 119–25; three-party coalition government demanded by, 44, 80, 81, 83, 99, 130; unwilling to agree to simple withdrawal-for-POWs deal, 155; US assurance of nonintervention in takeover of South, xii, 30, 35, 77–78, 80, 86, 97, 114–15, 117, 138–39, 146, 153, 162, 169, 175–76, 179, 180, 184, 190–91, 203, 251n13 (*see also* "decent interval" strategy); US presidential polling and, 84. *See also* "decent interval" strategy; Geneva Accords; Ho Chi Minh; North Vietnamese Army; settlement negotiations; Tho; Vietcong

Hanoi Radio, 7, 128, 152

Harris, Lou (pollster), 24, 64, 98, 196

Harvard Lampoon (periodical), 111–12

Heath, Edward R. G., 163

Henry Kissinger and the American Century (Suri), 189–92

Hersh, Seymour, 100

Hickenlooper, Bourke, 224n8

High Noon (film), 148

Hoang Duc Nha, 112, 116

Ho Chi Minh: death of, 72; French negotiations with, 9, 55, 70; Geneva Accords signed, 132–33; Nixon's

261

Ho Chi Minh (*continued*)
  reputation and, 163, 245–46n5;
  Nixon's secret ultimatum for, 70–74;
  political marginalization of, 231n11
Ho Chi Minh Trail: bombing of, 12, 13;
  China's concern about US offensives
  along, 32; infiltration via, 29, 73, 98,
  99, 100, 122, 153; "security guaran-
  tee" concerning, 94, 98–99, 100, 113,
  116, 117, 147, 153. *See also* Cambodia;
  Laos
Hornaday, Ann, 196
How Wars End (Rose), 203, 205
Huang Hua, 124–26
Humphrey, Hubert H.: on all-Indochina
  combat ban, 165; candidacy of, 98,
  140, 224n8; on Kissinger, 225n16;
  Nixon defended by, 47–48, 134;
  Reagan on, 37
Hung (Nguyen Tien Hung), 158,
  245n1
Hussein, Saddam, 208

Indochina: colonial legacy in, 102;
  French negotiations and withdrawal
  from, 55, 70; number of countries
  in, 113, 118, 147; US combat ban in,
  163–70. *See also* Cambodia; Geneva
  Accords; Laos; North Vietnam;
  South Vietnam
international supervision issue, 29, 56,
  105, 153. *See also* National Council of
  National Reconciliation and
  Concord
Iraqization, 201–2, 253n3. *See also*
  Vietnamization
Iraq Study Group, 170, 247n1
Iraq War (2003–2011): "Nixonian
  approach" in, 200–202; troop
  withdrawal recommended for, 170;
  unasked questions about, 206–9
Isaacson, Walter, 112
Islamic State insurgents, 201

Jamieson, Kathleen Hall, 241n2
JFK. *See* Kennedy, John F.

Johnson, Lyndon B. (LBJ): blamed for
  Vietnam, 37; bombing halted by, 70,
  98, 132, 140–41, 224n8; bombing
  offensives under (Rolling Thunder),
  154; Chennault Affair and, 140–41,
  149, 224n8; Pentagon under, 50;
  secret tapes of, 179; Tet Offensive
  and, 57; ultimatum to Hanoi, 71
Joint Chiefs of Staff: bombing restrictions dispute and, 67–68; on
  bombing under LBJ, 63; on South
  Vietnam's inability to survive, 14,
  48–49, 72, 198, 205
*Joint Chiefs of Staff and the War in
  Vietnam, 1971–1973, The* (study),
  153–54

Kaiser, Robert G., 250n1
Kalmbach, Herb, 108
Karnow, Stanley, 152
Katz, Mark N., 225n17
Kennedy, John F. (JFK): coup against
  Ngo Dinh Diem, 92, 138, 233n7;
  Cuba and, x, 37, 39–40, 168; election
  of, 98; family of, 192; mentioned,
  20; Nixon and, 2, 39–43; secret tapes
  of, 179; Vietnam, Laos, and, 37, 42,
  45–46
Kennedy, Robert (RFK), 43, 46, 48, 192
Kennedy, Rory, *Last Days* documen-
  tary of, 192–97
Kennedy, Ted: family of, 192; Nixon's
  war strategy critiqued by, xi, 45,
  49, 82–83, 138; on South Vietnam
  government, 43; on Vietnamization,
  xi, 197
Kerry, John, 206
Khe Sanh battle (1968), 60–61
Khmer Republic. *See* Cambodia
Khmer Rouge, 164
Khrushchev, Nikita, 39, 41
Kimball, Jeffrey: on Asselin's work,
  174, 175–76; on bombing's impact,
  228n33; book cover and, 212; on
  ceasefire and withdrawal, 249n2; on
  decent interval strategy, 173–74, 176,

178, 186, 189; on *FRUS* volumes, 177;
on Korean armistice, 70–71; on
Nixon's nuclear alert, 230n6
Kirk, Larry, 126
Kissinger, Henry A.: all-Indochina
combat ban and, 166, 169–70, 247n6,
251n11; biographers of, 189–92;
cartoon's depiction of, 111–12;
celebrity cult of, 36–37, 54–55; ego
of, 148; Ellsberg on, 53; Haldeman's
name and, 3, 215n6; Humphrey on,
225n16; legacy concerns of, 251n11;
media interaction of, 106–7, 129–31,
192, 195; media interaction on "Peace
Is at Hand" briefing, 133–35, 136;
Nixon's campaign stops and, 108–10;
Nixon's choice of, 2–3; office of, 11;
Paris Accords signed by, 159; secret
strategy for winning 1972 election
(*see* "decent interval" strategy);
secret talks with China, 30–36, 54,
78–81, 174–76, 190; secret talks with
Hanoi, 26–30, 54, 183; secret talks
with Soviets, 33, 70, 190; "security
guarantees" requested by, 94, 98–99,
100, 113, 116, 117, 130, 153; self-por-
trayed as above politics, 115–16,
129–31, 133; US aid cutoff to South,
threatened, 149–50, 154–57, 160, 176,
242n8, 244n32; Vietnam experts of
(*see* Halperin, Morton H.; Negroponte, John D.). *See also* "decent
interval" strategy; Paris Peace
Accords; settlement negotiations;
Thieu; Tho; Zhou Enlai
—topical comments: "behind the
trees" anecdote, 98; blame for
Cambodia's fall, 167–68; bombing
offensive, 59; bombing restrictions,
67; "decent interval" strategy,
188–89; destruction of liberals,
19–20, 30, 84, 104, 109; election
results (1972), 145–46; Geneva
Accords, 35; incursion into Laos,
13; in *Last Days* documentary, 192,
195; likely failure of bombing (1969),
72–73; media, 117; military vs.
political issues division, 28–29, 77,
79–80, 177–78; Nixon's key decisions, 182–85; Nixon's speeches, 11,
16, 18–19, 20; ping-pong diplomacy,
21; Son Tay prison raid, 8; South
Vietnamese elections, 87–88; Soviet
summit, 75–76; troop pullout and
POWs, 9; Vietnamese people, 119;
withdrawal deadline issue, 3–5, 7–8,
10–11
Korean War (1950–1953): China's
intervention in, 32; negotiations on,
70–71; Truman blamed for result in,
37, 38, 168; US residual troops in, 27
Kraft, Joe, 106

Laird, Melvin: aid-cutoff myth of,
200–201; all-Indochina combat ban
and, 165–66, 246n2, 247n6; on blame
for Cambodia's fall, 168; call for
bombing, 65; on Easter Offensive,
56; on Iraqization, 253n3; Nixon's
relationship with, 165–66; on
North's supplies, 62–63; plea to
release POWs, 6; on public response
to escalation, 72; on Vietnamization, 14
Laos: China's concern about US in, 32;
JFK's policy on, 42, 46; Nixon's
concern about collapse of, 29–30;
"security guarantee" concerning Ho
Chi Minh Trail and, 94, 98–99, 100,
113, 116, 117, 153; South Vietnamese
incursion in, 11, 12–14; Vietnam
settlement backed by, 123. *See also*
Ho Chi Minh Trail
Last Days in Vietnam (documentary),
192–97
LBJ. *See* Johnson, Lyndon B.
leaks, 12, 70, 178. *See also* National
Security Study Memorandum One;
Pentagon Papers
"leaving without losing," 48, 225n17
Le Duan, 231n11
Le Duc Tho. *See* Tho

LeMay, Curtis, 65–66
Lewis, C. S., 253n1
liberals: backstabbing myth affirmed by, 166–67; disarray of, 45; Nixon and Kissinger's plan to destroy, 19–20, 30, 84, 104, 109, 121, 123, 131, 167; Nixon's actual strategy misread by, xi–xii, 49–54, 101–3, 141–42, 151, 169, 172–73. *See also* Democrats
Life magazine, 45, 50
Lincoln, Abraham, 41, 143, 223n7
Lon Nol, 117
Lord, Winston, 30, 100
Los Angeles Times: on American POWs, 6; on China's opening, 36; on Khe Sanh battle, 60–61; on Nixon's troop withdrawals, 20–21; on North's troop withdrawals, 137; on Thieu's blowup, 139; on timing of Vietnam settlement and election, 107

MacArthur, Douglas, 32
MACV (Military Assistance Command, Vietnam), 198
Making of the President 1972, The (White), 143–44
Mansfield, Mike, 45, 134, 155–56
Mao Zedong, 33, 54, 81
Marder, Murrey, 196
Marshall, George, 38, 41
Martin, Graham, 196
McCain, John, III (POW and Senator), 6, 67
McCain, John, Jr. (Adm.), 67, 68, 71
McCarthy, Eugene J., 43, 48, 223n16
McCarthy, Joe, 38, 45–46
McGovern, George: on all-Indochina combat ban, 165; campaigning of, 82–83, 90, 127, 132, 139, 141–42; concession speech of, 145; on election as choice between "saving face or saving lives," 10, 48, 101–3, 127, 133, 141–42; on failure to sign Vietnam settlement, 139, 240n15; Nixon and Kissinger's plan to crush, 84, 104, 109, 121, 123, 131, 167; Nixon's actual strategy misread by, xi–xii, 82, 92, 101–3, 127, 141–42, 172–73; Nixon's linking of "surrender" to, 143, 147, 156, 167; Nixon's order to treat with contempt, 129; Nixon's view of, 103–4, 105, 144–45; POW/MIA bracelet of, 10; on Vietnamization, 52, 92; on Watergate break-in, 110
McNamara, Robert S., 40, 50
Medal of Honor, 18
media: Kissinger encouraged to bamboozle, 123; Kissinger's depiction of, 117; Nixon's deliberate leaking to, 70; video camera technology of, 131. *See also specific newspapers and television networks*
—topics covered: ARVN's advances, 69; China's opening, 36–37; "decent interval" strategy mentioned, 250n1; draft Vietnam settlement, 128–35; Easter Offensive, 56–58; Kissinger as celebrity, 36–37, 54–55; Kissinger cartoon, 111–12; Kissinger's ego, 148; Nixon and Fellows program, 126, 238n13; "Peace Is at Hand" briefing, 129–31; spring offensive results, 62; Thieu's objections, 126–27, 137–38, 139–40; timing of Vietnam settlement, 106–7; Vietnamization, 20–21; on Vietnam settlement nearly complete, 157–58
Military Assistance Command, Vietnam (MACV), 198
Miller Center, University of Virginia, xii, 179, 185, 187, 215n3, 253n2
Minh (Duong Van "Big" Minh), 140, 233n7
Minh, Ho Chi. *See* Ho Chi Minh
mining North's harbors (1972): context of, 64–67; continuation of, 89, 91, 129, 150; later reflections on, 70; as political cover for "decent interval" negotiations, 77–78; polling on, 55–56, 58, 64, 71–73, 118; rationales

for, 57–58; results of, 62–64, 69, 96, 97. *See also* bombing offensives
Moorer, Thomas, 67–68
Morris, Linda, 6
Mudd, Roger, 81
Muskie, Edmund, 51
myth of aid cutoff, 200–202. *See also* backstabbing myth

National Archives, 4, 172, 175, 189, 212. *See also* secret recordings
National Convocation of Lawyers to End the War, 45
National Council of National Reconciliation and Concord: Communists legitimated by presence on, 122; as provisional and powerless, 81, 83, 117, 125; Thieu's rejection of, 88–89, 117, 119, 127, 152; unanimity required for, 87–88, 105, 130, 132–33, 152
National League of Families of American Prisoners and Missing in Southeast Asia, 9–10, 107–10
National Press Club, 136–37
National Security Council (NSC), 50, 93. *See also* Kissinger, Henry A.
National Security Study Memorandum One (NSSM-1): disagreement in responses to, 15; leaking of, 63, 229n44; secrets in, 75; on South's survival (or not), 27, 94, 162, 183, 198
NBC Evening News: on China's opening, 36; on end of war, 26; Kissinger's "Peace Is at Hand" briefing covered, 131; Nixon's speech carried by, 2; on Thieu's demands, 126–27; on Vietnam settlement nearly complete, 158
NBC *Today Show*, 5
Negroponte, John D., 100, 122
Nessen, Ron, 194
Newsweek magazine, 54, 118–19, 131
New Yorker, 196
New York Times: on aid-cutoff threat, 157; on blame for fall of Cambodia, 169; on Cambodia and Laos bombing ban, 166; on China's opening, 36; on Easter Offensive, 57; on evacuations from Saigon, 195; on fighting after Vietnam settlement, 160; on Humphrey's defense of Nixon, 47; on Kissinger's briefing, 135, 136; on McGovern, 90, 141; on Nixon, 126; on Nixonian approach, 170–72, 203–4; on Nixon's troop withdrawals, 20–21; on spring offensive results, 62; on Thieu, 135–36, 139; on Vietnam settlement nearly complete, 157–58; on withdrawal of troops from Afghanistan, 254n6
New York Times Magazine, 37
Ngo Dinh Diem, 92, 138, 233n7
Nguyen, Lien-Hang T., 231n11
Nguyen Phu Duc, 116, 148–49, 155, 156
Nguyen Tien Hung, 158, 245n1
Nguyen Van Thieu. *See* Thieu
Nha (Hoang Duc Nha), 112, 116
Nichter, Luke, 185–87
Nixon, Richard M.: accusations against opponents embodied by, 39, 222n9; on appeasement, 38, 39, 222n9; call for bombing in 1960s, 64; campaigning of (1972), 103–4, 107–10, 122, 124, 128, 142–43; China's opening and visit of, 31, 36–37, 54–55; daily news summary for, 19, 21; fatal politics of, 202–6; game of politics of, 75–78, 85–86, 96–97, 145, 160–61, 168; JFK and, 2, 39–43; "kitchen debate" with Khrushchev, 39; legacy concerns of, 30, 44, 86, 96, 167–68, 186, 196–97; Madman Theory of, 163, 174, 245–46n5; McGovern's viewed as disloyal by, 102–3; Medal of Honor award presented by, 18; nuclear arms treaty of, 58, 64; Paris visit of, 95–96; post-Watergate suppression of evidence by, 189; secret strategy for winning 1972 election, ix–xii (*see also* "decent interval" strategy); "tooth episode"

Nixon, Richard M. (*continued*)
of, 144–45, 146; unasked questions about, 200–209. *See also* elections; *No More Vietnams*; secret recordings; speeches
—characteristics: anti-intellectual, anti-Ivy League, and anti-Semitic, 2; attitudes in secret tapes, 180–81; denial of responsibility, xii; foreboding felt, 143, 145–47; hospitality, 106; immune to irony, 92, 103, 189; leaks feared and used, 12, 70, 178; "miserable and paranoid," 122; music preferences, 144; negotiating style, 33; nonpartisan pose, 241n2; role model in, 95–96; self-puffery, 183
—war actions: "any means necessary" message, 70–74, 230n3; "decent interval" strategy denied by, 188, 189; decision making of, 170, 181–85; dispute with Air Force, 58–59, 67–69; four years needed to hide failure in Vietnam, 170–73; as model for twenty-first-century wars, 200–206, 253n1; nuclear threat, 70–71; secret ultimatum (1969), 70–74; strength of foreign policy touted, 19, 49–54, 65, 74, 77, 132, 134, 136–37. *See also* backstabbing myth; bombing offensive; ceasefire-in-place; "decent interval" strategy; evidence; fall of Saigon; mining North's harbors; POWs; public opinion polls; settlement negotiations; Vietnamization; withdrawal of troops

Nixon and Kissinger (Dallek), 251n13

Nixon's Vietnam War (Kimball), 173–74, 189

Nixon Tapes, The (Brinkley and Nichter), 185–87

No More Vietnams (Nixon): on ARVN, 61; backstabbing myth in, ix, xii, 165–66, 171–72, 191; on bombing and mining, 55, 64, 76; on decent interval strategy, 189; on Hanoi's supply lines, 63

No Peace, No Honor (Berman), 172–73

Norodom Sihanouk (prince), 163–64

North Vietnam: call for bombing of (1960s), 64–66; Chinese troop support for, 63; nationalism of, 72; overestimating hardships of, 96; US bombing and mining of (*see* bombing offensives; mining North's harbors); US raid on Son Tay prison in, 8, 161–62. *See also* Hanoi government

North Vietnamese Army (NVA): Easter Offensive of (1972), 56–62, 69, 75–76, 102, 205; face-saving withdrawal of some troops from South, 94–95, 120–21, 125, 147; March 1975 offensive of, 193; number of troops, 56–57, 227n7; one-to-one demobilization of South and (proposed), 147; Saigon's forces vs., 27, 53–54; shift deeper into Cambodia, 12; supply lines cut, 13–14 (*see also* Ho Chi Minh Trail); total troop withdrawal from South demanded by Thieu, 136–40; treaty as legitimating presence in South, 122; troops remaining in South, 106, 113, 114, 116–17, 118, 183, 193, 198–99, 200, 232n4

Novak, Robert, 70

November Group, 101

NSSM-1. *See* National Security Study Memorandum One

nuclear weapons: Kissinger's secret negotiations on, 32–33; mythical missile gap in, 40–41; Nixon's implied threat of, 70–71, 245–46n5; Reagan and fear of, 20; US-Soviet treaty (SALT) on, 58, 64

NVA. *See* North Vietnamese Army

Obama, Barack: on troop withdrawals from Afghanistan, 208–9, 254n6; urged to use Nixonian approach, 202–6, 253n1

O'Donnell, Kenneth P., 45–46
Office of International Security (Pentagon), 50
Operation Linebacker I (May 8) and II (Christmas). *See under* bombing offensives
Operation Pocket Money. *See* mining North's harbors

Packer, George, 196
Palace File, The (Nguyen and Schecter), 158, 245n1
Paris Peace Accords (1973): denial of real implications of terms, 133, 134; expected violations of, 35, 77–78, 80, 114–15, 117, 138–39, 162, 169, 175 (*see also* fall of Saigon); Geneva Accords compared with, 132–33; implications of US bombing after signing, 161–63; Nixon's high approval ratings after signing, 159–60, 244n3, 245n5; North's willingness to sign, 190–91. *See also* settlement negotiations
Passport (review journal), 178
peace: China's opening in context of, 36; LBJ's attempts, 98; Nixon's publicly declared desire for, 17–18, 25, 124, 136, 142–43, 158, 159, 178
—"Peace Is at Hand" briefing: Democrats' responses to, 132–35; Kissinger's role in, 129–31; Right's response to, 133–34
—"peace with honor": context of, 49, 51–52; Nixon's speeches emphasizing, 86–87, 109–10
Pelosi, Nancy, 207
Pentagon Papers, 46, 50, 53. *See also* US Dept. of Defense
People's Republic of China. *See* China
Peterson, Pete, 119
Pham Van Dong, 118–19
ping-pong diplomacy, 21, 31–33
Polo I. *See* China: Kissinger's secret talks with
post hoc, ergo propter hoc (fallacy), 69, 230n4

Powerful and Brutal Weapons (S. Randolph), 61, 232n4
Powers, Dave, 46
POW/MIA campaign, 9–10, 107–8
POWs (prisoners of war): dilemmas concerning, after signing Vietnam settlement, 161–63; families and organizations supporting, 9–10, 107–10; Hanoi's condition for release of, 7–10, 25–26, 27, 52, 62, 91, 99, 108, 154, 161–62, 184, 186, 205, 217n38; Hanoi's treatment of, 6–8, 24; McGovern on, 102; Nixon's actual comments on timing of release, 7–8, 91; Nixon's publicly declared concerns for, 8–10, 17–18, 25–26, 151; North's commitment to releasing, 151; "Peace Is at Hand" briefing and wives of, 134; release according to Paris Accords, 159; released POWs' views on reparations for Hanoi, 162–63; withdrawal-for-POWs deal proposed in Congress, 154–56. *See also* casualties, US; "decent interval" strategy
Presley, Elvis, 4
Price of Power, The (Hersh), 100
prisoners of war. *See* POWs
public opinion polls: bombing and mining to end war, 55–56, 58, 64, 69, 71–73, 118, 124, 150, 159, 244n3, 245n5; Cambodian incursion, 12; ceasefire-in-place, 118; Communist China, 31–32; Communist takeover of South Vietnam, 24; Ford's aid request, 196; Nixon's approval ratings, 22, 49, 64, 159–60, 244n3, 245n5; Nixon vs. McGovern, 87, 98, 121, 122; potential withdrawal deadlines, 23–24; POW rescue attempts, 8; questions not asked, 24; secret, 55–56; troop withdrawals, 2, 4, 10, 20–23; Vietnam exit strategies, 48, 55–56, 58

Radio Hanoi, 7, 128, 152
Randolph, John, 60–61

Randolph, Stephen, 61, 228n38, 232n4
Rather, Dan, 21
Reagan, Ronald: on bombing, 66, 150; on China's opening, 37, 38–39; Ted Kennedy denounced by, 46–47; Kissinger praised by, 133–34; on Korean War, 168; Nixon and, 19–20, 64, 86; POW/MIA bracelet of, 9–10; on Vietnam War, 220n8
recriminations, 16–17, 20, 49
Republican Coordinating Committee, 65
Republican National Committee, 55–56
Republicans: blamed for losing countries to Communism, x, 48; bright future of, 19–20; call for bombing in 1960s, 64–66; on China's opening, 37; convention of (1972), 86–87; election results (1946), 168; election results (1972), 145; on losing the war, xii; Nixon's hope for triumph of, 19–20, 30; Nixon's strategy supported by, 46–47, 136–37; on "Peace Is at Hand" briefing, 134; strength of foreign policy touted, 49–54, 134, 136–37, 207
Republic of South Vietnam Armed Forces. *See* Army of the Republic of [South] Vietnam
"residual troops" idea, 27
Reston, James "Scotty," 135
Rockefeller, Nelson, 31, 122, 136–37
Rodgers, Richard, 144
Rogers, William, 6, 97, 159
Rolling Thunder campaign. *See* Johnson, Lyndon B., bombing offenses under
Romney, Mitt, 208
Roosevelt, Franklin D., 37, 144, 168, 179, 224n8
Roosevelt, Theodore, 33, 183
Rose, Gideon, 202–6, 253n1
Rosenthal, A. M., 136
Rusk, Dean, 46
Russell, Richard, 65

Russia. *See* Soviet Union
RVNAF. *See* Army of the Republic of [South] Vietnam
Ryan, John, 67–69

SAC (Strategic Air Command), 71
Saigon government: expendability of, 160–61; Ford's last-minute aid requests for, 193–96; hopes to keep US troops in country, 88–89; infantilization of, 91, 114–15; liberals' misreading of Nixon's support for, xi–xii, 49–54, 101–3, 141–42, 151, 169, 172–73; Nixon's secret go-between with (1968), 140–41, 149, 224n8; Nixon's threats to, 116, 117, 147, 148–50, 158, 242n8; POWs held by, 111, 112–13, 148, 151; "security guarantees" for, 94, 98–99, 100, 113, 116, 117, 130, 153; settlement terms as eventual destruction of, 7–8, 49, 77–78, 85–86, 88, 89–95, 96–97, 112, 113–14, 115–16, 175, 177, 180, 183, 185–86, 196–97; US aid cutoff threatened, 149–50, 154–57, 160, 176, 242n8, 244n32; US aid for (1975), 200–201. *See also* Army of the Republic of [South] Vietnam; "decent interval" strategy; fall of Saigon; Geneva Accords; settlement negotiations; South Vietnam; Thieu
Sainteny, Jean, 70, 74, 230n3
SALT. *See* nuclear weapons
Schecter, Jerrold L., 158, 245n1
Schlesinger, Arthur M., Jr., 39–40
Schmitz, David F., 245n5
Scott, Hugh, 22–23, 25
Screwtape Letters, The (Lewis), 253n1
secret recordings: absent from "Western White House" (San Clemente), 165, 167; dates covered by, ix; declassification of, 172; as "gift that keeps on giving," 209; microphone placement, 1; Miller Center's transcriptions noted, xii, 179–80,

185, 187, 215n3, 253n2; voice-activated system, 3. *See also* "decent interval" strategy; evidence
Secrets (Ellsberg), 53
Secret Service, 1
self-determination, 128, 129, 130, 159
settlement negotiations
—1968 (Johnson administration): Nixon's scuttling of, 140–41, 149, 224n8; US aid cutoff threatened, 157
—1969–1972 (Nixon administration): "behind the trees" anecdote about, 98; denial of secret side agreements in, 131; earlier secret meetings with China concerning, 35–36; final deal desired after election, 122–26; French delegate for Nixon in, 70, 74, 230n3; Hanoi's handling of draft agreement, 121, 124, 128–30; Hanoi's October 31 deadline for signing, 98, 104, 125, 128, 136–37; Hanoi's settlement proposal (Oct. 8, 1972), 99–100, 184–85, 250n9; hopes for pending settlement, 89–97; Nixon and Kissinger's strategizing about, 83–86, 103–10; Nixon's China visit and, 54–55; Nixon's key decision points in, 181–85; Nixon's public vs. actual stance in, 86–87; Nixon's secret ultimatum (1969), 70–74; potential blowups and falling apart of, 107, 111, 112–13, 119–26; South and war lost in, 196–97; Thieu's demands in, 136–40; Thieu's rejection and stalling on, 106, 117–25, 126–27; timing of signing deal, 106–7; US assurance of nonintervention in North's takeover of South, xii, 30, 35, 77–78, 80, 86, 97, 114–15, 117, 138–39, 146, 153, 162, 169, 175–76, 179, 180, 184, 190–91, 203, 251n13; withdrawal of North's troops from South demanded, 117, 125, 127, 136–40, 183. *See also* bombing offensives; "decent interval" strategy; mining North's harbors
—late 1972–1973 (Nixon administration): Hanoi's commitment to, 151–52, 243n6; Kissinger's list of Hanoi's concessions from Nov. 7 to Jan. 27, 152–53; promises and threats to North and South, 147–50, 242n8; text completed, 158; US aid cutoff to South, threatened, 149–50, 154–57, 160, 176, 242n8, 244n32; US bombing continued during, 150–54; withdrawal-for-POWs deal proposed in Congress, 154–56. *See also* Paris Peace Accords
—specific issues: free elections in South, x, xi, 26, 55, 56, 87–88, 89, 99, 105, 117–18, 130, 132, 146, 233n11; Hanoi's condition for release of POWs, 7–10, 25–26, 27, 52, 62, 91, 99, 108, 154, 161–62, 184, 186, 205, 217n38; international supervision, 29, 56, 105, 153; Kissinger and Thieu's negotiations concerning, 111–16; military vs. political euphemism for, 28–29, 77, 79–80, 177–78; mutual withdrawal of troops proposed, 27, 73, 147, 183, 249n2; Nixon and Kissinger's key issues, 26–30, 99, 135, 151, 152–53, 175; Nixon's concern about South's collapse before 1972 election, 3–4, 7, 28–30, 91–92, 108, 124, 176, 185, 203–5; POWs held by Saigon, 111, 112–13, 148, 151; "security guarantees" requested from Hanoi, 94, 98–99, 100, 113, 116, 117, 130, 153; Thieu's resignation (later dropped), 81, 84, 99, 105, 183–84, 249n5; total withdrawal of North's troops from South demanded, 117, 125, 127, 136–40, 183; US reparations, 99, 105, 130, 162–63, 235n6; US resupply of armaments for South, 99, 105, 111, 112, 152. *See also* ceasefire-in-place; coalition government; "decent interval" strategy; National Council of National Reconciliation and Concord; POWs; Thieu; Tho; withdrawal of troops

269

Shriver, Sargent, xi, 83, 138–39, 192, 197
Sihanouk (Prince Norodom Sihanouk), 163–64
Silent Majority speech (November 3, 1969), 49, 74–75
Sloan, Hugh, 128
Snepp, Frank, 187–88, 193, 194–95
Son Tay prison raid, 8, 161–62
Sorley, Lewis, 59, 197–200
South Vietnam: blame for loss of, 30, 42–43, 44; "chance" to survive as a free people, 15–16, 120–22, 180–81, 197–98, 206; "Communist takeover" in polls about, 24; coup (1963) in, 92, 138, 233n7; election rigged in (1971), 140; evacuations from, 193–97; expected Communist takeover of, 7–8, 11, 49, 77–78, 85–86, 88, 89–95, 96–97, 112, 113–16, 138–39, 147; hopes for free elections in, x, xi, 26, 55, 56, 87–88, 99, 105, 130, 132, 146, 233n11; JFK's military buildup in, 42–43; Kissinger cartoon in newspaper, 111–12; Kissinger's reflections on, 123; National Assembly rally in, 136; National Day events in, 138, 140; Nixon's public refusal to impose Communist government on, 43–44, 77–78, 103, 110, 128, 129, 130, 159; "no one will give a damn about," 85–86; North's infiltration of, 29, 73, 98, 99, 100, 122, 153; North's troops remaining in South, 106, 113, 114, 116–17, 118, 183, 193, 198–99, 200, 232n4; US aid for (1975), 200–201. *See also* Army of the Republic of [South] Vietnam; ceasefire-in-place; fall of Saigon; Saigon government; settlement negotiations
Soviet Union: ambassador of (*see* Dobrynin, Anatoly); attitude toward Nixon, 74, 76; attitude toward US, 97; hostility with China, 31, 34; implied threats to, 70–71; intervention in war, 76; interventions in Africa, 190;
Kissinger's negotiations and Soviet role in "decent interval" strategy, 30, 118, 162, 190, 205; Kissinger's secret talks with, 33, 70, 190; Nixon's China visit and, 55; North supported by, 63; nuclear capabilities of, 40–41; settlement negotiations and, 90, 119–20, 122–23; US nuclear treaty with, 58, 64; US-Soviet summit planned (1972), 75–76
speeches (Nixon): "ad libbed" speech ending for, 17–20, 21–22; on ceasefire-in-place and mutual withdrawal, 249n2; "clearest choice," 142–43; peace desired, 17–18, 25, 124, 136, 142–43, 158, 159, 178; "peace with honor" emphasized in, 86–87, 109–10; polls and approval ratings after, 20–23, 49, 159–60; to POWs' families, 109–10; practicing for, 11; "tooth episode" and, 144–45, 146. *See also under* elections
—specific: April 7 (1971), 1–2, 5–6, 8, 13–16, 17–22; April 26 (1972), 57; January 23 (Paris Accords signing, 1973), 159–60; May 8 (1972), 55–58, 61–62; November 3 (Silent Majority, 1969), 49, 74–75
Spirit of '76 (president's plane), 143–44
stabbed-in-the-back myth. *See* backstabbing myth
Stalin, Josef, 37
Stennis, John C., 149, 156–58, 160, 176, 244n32
Stevenson, Adlai, 222n9
Stockdale, Sybil, 6, 8
Strategic Air Command (SAC), 71
Streithorst, Tom, 126
Suri, Jeremy, 189–92
surrender: coalition government as, 44, 98, 119, 127, 197; "decent interval" strategy as, 44, 150, 162, 196–97; Nixon's linking of McGovern to, 143, 147, 156, 167; Vietnamization as negotiating, xi–xii; Vietnam settlement as, 138

Taiwan: China's opening as threat to, 36
Tangled Web, A (Bundy), 189
Taylor, Karl, 18
Taylor, Karl, Jr., 18
Taylor, Kevin, 18, 21, 22
Taylor, Shirley (Mrs. Karl), 18, 21
Tet Offensive (1968), 57
Thailand: government of, 96; US air force in, 28; US relationship with, 169
Thieu (Nguyen Van Thieu): blowup in 1968, 97–98, 132, 140–41; blowup in 1972, 136–40; "chance" of survival, 15–16, 120–22, 180–81, 197–98, 206; coup considered against, 92–93; ground offensive halted by, 13; hopes to keep US troops in country, 88–89; Kissinger's view of, 111–16; media's questions about, 131; needed in place for successful "decent interval" strategy, 93, 100, 105–6, 115–16, 122; Nixon's attitude toward, 84, 85; Nixon's discussion of how to handle, 104–6, 107; Nixon's Inauguration Day ultimatum for, 158; on possibility of US bombing after signing Vietnam settlement, 161; refusal to sign declared in vague terms, 135–36; resignation demanded then dropped as issue, 81, 84, 99, 105, 183–84, 249n5; settlement rejected by, 106, 117–25, 126–27; settlement terms as eventual destruction of, 77–78, 89–90, 92, 93, 112, 113–14, 115–16, 118, 175, 177, 185–86; US aid cutoff threatened, 149–50, 154–57, 160, 176, 242n8, 244n32; US liberals' misreading of Nixon's relation to, xi–xii, 49–54, 101–3, 141–42, 151; withdrawal of North's troops from South demanded, 117, 125, 127, 136–40, 183
Tho (Le Duc Tho): commitment to Vietnam settlement, 151, 243n6; "decent interval" strategy and, 87, 180; Paris Accords signed by, 159; pressures on, 90; settlement proposal of (Oct. 8, 1972), 99–100, 184–85, 250n9; on Thieu's proposed changes, 147–48; Thieu's resignation demanded, 81; on three-party coalition government, 83. *See also* Hanoi government; settlement negotiations
Thompson, Llewellyn, 74
Thompson, Robert, 72, 73
Thunderbolt (Sorley), 59, 199
Time magazine, 36–37, 54
Toledo Blade (newspaper), 238n13
Toth, Bob, 107, 137
Tran Van Lam, 116
Truman, Harry S.: alleged missile gap left by, 41; blamed for countries lost to Communism, 37, 168; China and, 32, 38; secret tapes of, 179; Vietnam and, 53

United Republicans of California, 37
United States: alleged effects of all-Indochina combat ban on reputation, 169–70, 251n11; geopolitical credibility forfeited, 190–91; Nixon's concern about others' perceptions of, 96–97; Soviet nuclear treaty with, 58, 64; US-Soviet summit planned (1972), 75–76
University of Virginia, Miller Center, xii, 179, 185, 187, 215n3, 253n2
US Air Force: accused of undercutting Nixon, 58–59, 67–69; power of, 61, 64, 71; restrictions on, 65–66, 67–68; Thailand bases, 28. *See also* Abrams, Creighton W.; bombing offensives
US Congress: aid for South Vietnam from (1975), 200–201; all-Indochina combat ban by, 163–70; continuing resolution for budget, 166; facts vs. accusers in, 38–39; Ford's last-minute aid requests from, 193–96; Nixon's opinion of leaders in, 20; Nixon's threat of aid cutoff in name of, 149–50, 154–57, 160, 176, 242n8,

US Congress (*continued*)
244n32; Rose's blame for defeat on, 205; threatened recriminations for withdrawal decision and, 16–17; troops prohibited from Ho Chi Minh Trail, 13; withdrawal deadline issue and, 9–10, 47; withdrawal-for-POWs deal proposed, 154–56. *See also* backstabbing myth

US Dept. of Defense: advisers in, 50; blamed for failures in Vietnam, 97; on bombing under LBJ, 63; Pentagon Papers of, 46, 50, 53; plea to release POWs, 6; on South Vietnam's inability to survive, 48–49, 72, 198, 205. *See also* American troops; Defense Intelligence Agency; US Air Force

US Dept. of State: on bombing under LBJ, 63; China White Paper of, 38–39; plea to release POWs, 6; on South Vietnam's inability to survive, 48–49, 72, 205

US Senate: Vietnam hearings of, 155–56

US-Soviet treaty (SALT), 58, 64

USSR. *See* Soviet Union

Veterans Day ceremony, 124
Victory at Sea (documentary), 144
Vietcong (VC): Saigon's forces vs., 27, 53–54, 183. *See also* Hanoi government
Vietnam Chronicles (Sorley), 59
Vietnamization: alleged success of, 13–14, 25–26, 72, 201; as delay tactic, xi, 45, 49, 82–83; as fraud and failure, xi, 14–16, 27, 43, 72, 180–81, 197, 203, 204–5; historical misreading of, 170–71; media response to, 20–21; Nixon's actual views on, x, 14–16, 182–83, 203; Nixon's public position on, 13–14, 15, 25–26, 49, 61, 74–75, 171, 180–81; "of the peace," 136; as perpetuating the war, 52; spring offensive as test of, 59–62; withdrawal of troops in context of, 27, 51–52. *See also* Army of the Republic of [South] Vietnam; fall of Saigon; Iraqization; National Security Study Memorandum One

Vietnam War: actual events vs. conventional story of end, 170–73; China's role in, 32, 62–63, 76–77, 220n8; decision to prolong, 3–5, 7–8; Easter Offensive (1972) in, 56–62, 69, 75–76, 102, 205; end of war possible (1969), 44; escalations of US involvement, 53–54; heroism in, 18; iconic images of, 102, 105; JFK's military buildup in, 42–43; Khe Sanh battle (1968), 60–61; LBJ's bombing halt in, 224n8; liberals' belief in Nixon's perpetual war in, 49–54; Nixon's actual vs. depicted actions in, 146–47; Nixon's exit strategy (*see* "decent interval" strategy); Nixon's key decision points in, 181–85; pacification in, 53; presidential campaigns' stories on (1972), 83–84, 90; public view of outcome, 138–39; recriminations for loss in, 16–17, 20, 49; "search and destroy" vs. "clear and hold" strategies in, 200, 207; Sorley's declaration of winning, 197–200; two paths to defeat in, 142–43; ultimate cost of, 86, 196–97, 200–209; unasked questions about, 205–6; as unwinnable, 189, 191–92, 199. *See also* American troops; bombing offensives; casualties; Hanoi government; mining North's harbors; Paris Peace Accords; POWs; Saigon government; settlement negotiations; South Vietnam; withdrawal of troops

Vietnam War Files, The (Kimball), 173–74, 176

Wallace, George, 19
Wall Street Journal, 139–40, 166

Walters, Barbara, 5
Warnke, Paul C., 50–51, 52
Washington Post: aid-cutoff myth in, 201; on China's opening, 36; on Ford's aid request, 196; on Nixon and Fellows program, 126; on Nixon's announced troop withdrawals, 20–21; on Nixon's strategy, 51–52; on POW release and Vietnamization linkage, 26; on Thieu, 140; on Vietnam settlement and election, 92; Watergate exposé in, 108, 128
Washington Star-News, 106
Watergate: exposé on, 108, 128; McGovern on, 110; Nixon's cover-up of, 178–79; pending hearings on, 166–67; settlement negotiations as top news over, 132, 137
Westen, Drew, xi
Westmoreland, William, 200
White, Theodore H., 143–44
White House: chief of staff (*see* Haldeman, H. R. "Bob"); Daily Diary referenced, 165; Fellows program of, 126; Kissinger's press briefing at, 130–31; Nixon's election day return to, 144; Nixon's favorite place in, 18; Watergate involvement denied, 128
Whitten, Paul, 164
withdrawal of troops: from Afghanistan, 254n6; all US forces included in, 62, 87, 161–63; deadline discussed, 10–11, 50; decision to time withdrawal to 1972 election, x, 3–5, 7–8, 10–11, 15–16, 25–26, 28, 48–49, 179–81, 182–83, 186, 203; Democrats and what-ifs in, 47–49; as Hanoi's condition for release of POWs, 7–10, 25–26, 27, 52, 62, 91, 99, 108, 154, 161–62, 184, 186, 205, 217n38; increases in, 2, 5–6, 9, 15, 20–21, 57; killing public pressure for Dec. 1971 deadline of, 11, 24–26; last combat team headed home announced, 86; mutual, 27, 73, 147, 183, 249n2; Nixon's deniability for Communist takeover after, 180; Nixon's televised speech on, 1–2, 5–6, 8, 15–16, 20–21; in Paris Accords, 105, 159; politicians' link of 1972 election to timing of, xi, 22–23, 45–47, 49, 82; polls on Dec. 1971 date, 23–24; refusal to set, 5–6, 10–11, 16–17; "residual troops" idea and, 27; Vietnamization in context of, 51–52; withdrawal-for-POWs deal proposed in Congress, 154–56; Zhou's view on, 55. *See also* ceasefire-in-place; "decent interval" strategy; POWs; settlement negotiations
Woodward, Bob, 108, 128, 209
World Affairs Council, 138–39

Yates, Sidney, 164
Years of Upheaval (Kissinger), 169–70

Zhou Enlai: charisma of, 34–35; on Charles de Gaulle, 95; "decent interval" strategy discussions of, 81, 85, 191; intervention in Korean War, 31–32; Kissinger's secret talks with, 31, 34–36, 78–81, 175–76; mentioned, 125; Nixon's visit with, 54–55
Ziegler, Ron, 21–22, 128
Zorza, Victor, 140